THE MATH GNOME AND COMMON CORE 4

Written by
Diane Taylor, Michele Meyer, and Kelly Santora

Editor: Christie Weltz
Cover Illustrator: Kammy Peyton
Designer/Production: Isabelle Gioffredi/Kammy Peyton
Art Director: Moonhee Pak
Project Director: Stacey Faulkner

© 2014 Creative Teaching Press Inc., Cypress, CA 90630
Reproduction of activities in any manner for use in the classroom and not for commercial sale is permissible.
Reproduction of these materials in any manner, in whole or in part, for an entire school or for a school system is strictly prohibited.

DEDICATION

We would like to dedicate this book to all teachers out there who, like us, love to learn and live to teach!

ACKNOWLEDGMENTS

First, we would like to thank our families. We would never be where we are today if not for their love and support. They have always believed in us, and their faith has made us who we are today. We would also like to thank Keith Wing. He has been more than a principal to us. He has been a friend, mentor, and motivator. We are extremely grateful to Loren Penman and Rachael Greene for sharing their wealth of knowledge with us. Finally, we would like to thank our students. They have helped shape this program and are the reason we love doing what we do… teach!

We would like to acknowledge all of the best practices books, theories, and guides that have inspired us and pushed us toward becoming the best teachers we can be. The ideas and instructional methods that we have learned about and used in our classrooms have helped us to create and mold the Math GNOMe and Common Core 4 into what they are today.

We would like to especially acknowledge extraordinary educators such as Harry Wong, The 2 Sisters (Joan Moser and Gail Boushey), Debbie Diller, and Lucy Calkins for paving the way toward the current focus on independent, student-led, teacher-facilitated learning. We truly believe the key to creating lifelong learners is through empowering students to take charge of their education.

FOREWORD

Over the past two years, it has been my distinct honor as principal of Byron-Bergen Elementary School in Western New York to work closely with Kelly Santora, Michele Meyer, and Diane Taylor. These three individuals have truly taken the Mathematics Common Core State Standards to another level in their own classrooms, and I am glad to see that they want to share this knowledge with you. Instruction in mathematics does not come naturally to many of us who work at the elementary or primary levels. As a former fourth grade teacher, I remember looking at the clock and dreading 1:00, when it was time to teach math. I remember muddling through lessons, which I'm sure did no favors to students in my class. I always yearned for a more robust and engaging way to teach mathematics. And quite frankly, until I had observed Kelly, Michele, and Diane in action, I had never seen such an effective model for mathematics instruction.

The Math GNOMe and Common Core 4 components drive at the true intent of the Common Core State Standards. Rather than focusing on mathematics instruction "a mile wide and an inch deep," the Math GNOMe and Common Core 4 focus on key areas of instruction and drill deep for rigorous and more relevant instruction in specific targeted areas. But perhaps the greatest outcome of using the Math GNOMe and Common Core 4 is the high level of engagement it brings to mathematics instruction. This is something that is often lacking, particularly with teachers who do not feel as comfortable teaching math as they do other subjects. The Math GNOMe and Common Core 4 bring a student-centered approach to math. Students are actively engaged and take responsibility for their own growth. Furthermore, the Teacher Conference Binder helps the teacher more efficiently differentiate instruction and, in essence, create individualized learning plans for each student.

You have already read this foreword, which tells me that you are interested in learning ways to enhance mathematics instruction. Having seen the program firsthand, I can tell you that by using the Math GNOMe and Common Core 4, you will have a succinct plan for teaching mathematics, you will have a solid foundation for creating highly engaging lessons, and you will be able to individualize your instruction to each student in your classroom.

Keith Wing

Elementary Principal
Byron-Bergen Central School District

TABLE OF CONTENTS

Introduction	5

Chapter 1: The Math GNOMe — 10
- About the Math GNOMe — 10
- Implementing the Math GNOMe — 11

Chapter 2: The Common Core 4 — 23
- About the Common Core 4 — 23
- Implementing the Common Core 4 — 23
  - Math Fluency — 24
  - Math Games — 31
  - Mathematical Practice — 37
  - Technology — 43
- Teacher Conferences — 50

Chapter 3: Preparing Your Classroom — 57
- Bulletin Boards — 57
- Math Games Binder — 58
- Math Boxes — 58
- Manipulatives — 59

Chapter 4: Putting It All Together — 60

Chapter 5: Assessment — 67

Chapter 6: Home Connection — 69

- Common Core State Standards for Mathematical Practice — 76
- Common Core State Standards for Fourth Grade — 77
- References — 80

INTRODUCTION

Did you panic when you heard that your state was adopting the Common Core Standards?

If you are anything like us, you probably found yourself moving through the stages of panic, fear, curiosity, and, eventually, acceptance. At first we thought we were going to have to completely overhaul our entire curriculum and instruction. However, after exploring the Common Core Standards through professional development provided by our district, as well as on our own initiative, we came to the realization that we already had all of the tools needed to become a Common Core classroom.

After a few planning periods and lunch chats, we finally experienced our "aha" moment. What we would need to do is change the way we think about math instruction. The shift would come in the presentation of our lessons and in the way we ask our students to respond to math. The Common Core Standards mention shifts in the direction in which math is approached, so we needed to rethink the way in which we incorporated these standards into our math lessons and the level in which we allowed students to experience the content. The Common Core Standards take a constructivist approach to math learning. Therefore, students should be using their math knowledge to solve novel, real-world problems.

> **What we would need to do is change the way we think about math instruction.**

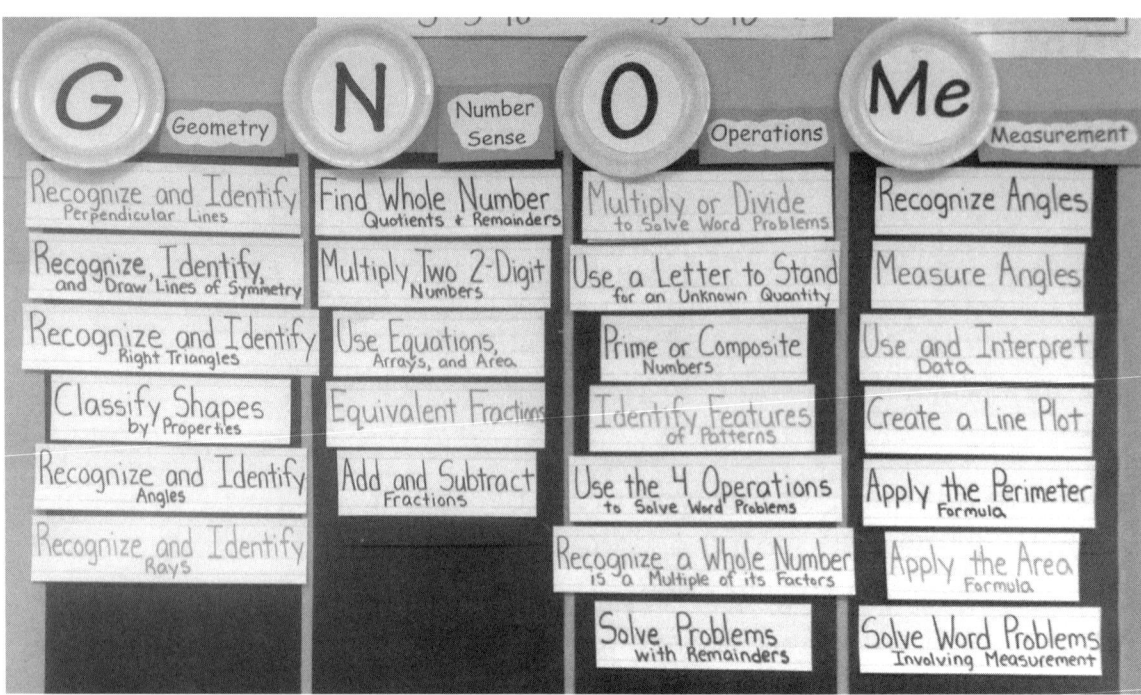

We really wanted our students to take ownership of their learning and be aware of the high expectations that are placed on them. In order to do this in a student-friendly way, we came up with the acronym GNOMe, which stands for geometry, number sense, operations, and measurement. These are the domains that the Common Core Standards use in organizing the standards. We then rewrote the standards in short, student-friendly phrases by highlighting the core math concepts addressed.

We couldn't think of a better way for students to grasp what they were expected to learn than by displaying the concepts to refer to throughout the day. Throughout the school year, as we introduce and teach the standards, we place each standard's core concepts under the corresponding domain (see Chapter 1 for more details). This is also a great way for us to make sure we are consistently teaching to the standards and are able to refer back to standards that were previously taught.

Next, we turned our attention to the setup of our math time. With the current emphasis on differentiation and individualized learning plans, we wanted to make sure our math block allowed us to work both one-on-one with individual students and in small groups with students who are all working on the same math skills. We decided that a mini-lesson followed by independent practice would work best. Marzano, Pickering, and Pollock (2001) discuss the importance of helping students set high expectations and objectives for themselves as well as having teachers provide immediate feedback. The independent practice time allows teachers to conduct a conference and provide meaningful feedback to individual students while the rest of the class is working on a task.

This format posed another obstacle. How do we meet with students individually and not lose meaningful instruction time with those working independently? We did not want to assign "busy work" to our students. We want our students to be engaged in meaningful activities that support their learning. We also want to spend most of our time working with students, not making copies and grading papers. This led to the birth of the Common Core 4, which is based on the Common Core State Standards for Mathematical Practice (page 76). We decided that there are four main activities that students should experience on a regular, consistent basis

> **How do we meet with students individually and not lose meaningful instruction time with those working independently?**

in order to support their math learning and growth. The Common Core 4 is composed of four kinds of activities: math fluency, math games, mathematical practice, and technology. (See Chapter 2 for more details.)

We also agreed that student choice needed to play a role in our math instruction. There is much research that shows students are more invested in their learning when they are interested in it. Katz and Assor (2007) discuss the impact that student choice can have on motivation, well-being, and achievement. What better way to appeal to the interest of students than by giving them choice when it comes to their learning. We decided to give students the opportunity to choose which one of the Common Core 4 they would like to work on and when.

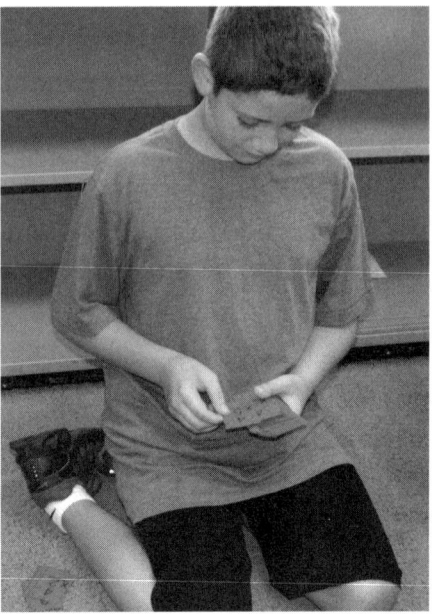

Putting it all together, our math day eventually ended up looking like this: We teach a mini-lesson about a concept on the GNOMe board. That is followed by students choosing which one of the Common Core 4 activities they would like to work on independently. While students work independently, we meet with individuals or small groups to reinforce skills. We repeat this two to three times within our math block, resulting in two to three mini-lessons and two to three rounds of the Common Core 4 every day.

> **The Common Core 4 is composed of math fluency, math games, mathematical practice, and technology.**

While we chose for the Common Core 4 to be the independent practice part of our math block, there are many other great math programs out there that would work equally as well in its place. If your school has a math series that you are required to use, the activities and lessons found in your series could take the place of the independent practice. Chapter 4 provides details and examples of how to set up your math block using the Math GNOMe with the Common Core 4, as well as with a variety of other options in place of the Common Core 4.

The Math GNOMe and Common Core 4 helped to align our instruction to the Common Core Standards with a simple shift in how we think about and present math. It is evident to anyone who walks into our classroom what standards we have taught and what standards students are working on. Our students are able to discuss math concepts and explain why

they are learning a particular skill. We, as well as students, are able to refer back to the Math GNOMe board throughout the school year. The Math GNOMe has focused our attention on the expectations the standards place on our students, while the Common Core 4 allows us time to work with students individually in areas where they are struggling. It is amazing to be able to find the time to work one-on-one with students on a regular basis. Because of this individualized attention and ongoing formative assessment, students' growth is astounding.

The Math GNOMe and Common Core 4 work in conjunction with one another to allow you to have a truly Common Core classroom that addresses the six shifts in mathematics instruction—focus, coherence, fluency, deep understanding, application, and dual intensity. Through the Math GNOMe mini-lessons, teachers are able to provide instruction that has focus, coherence, and depth for greater understanding, while the Common Core 4 provides students with the opportunity to achieve fluency, application, and dual intensity. Your students will participate in independent, self-selected math activities while allowing you to provide individual and small-group instruction and assessment through conferences.

The Math GNOMe and Common Core 4 helped to align our instruction to the Common Core Standards with a simple shift in how we think about and present math.

CHAPTER 1

ABOUT THE MATH GNOMe

The Math GNOMe

The Math GNOMe is a management system designed to help teachers easily incorporate Common Core Math Standards into the classroom.

To make the Common Core Standards student-friendly, we created the acronym GNOMe, which stands for geometry, number sense, operations, and measurement. These are the four domains, or categories, that span all grade levels in the Common Core Standards.

After we became more familiar with the Common Core Standards, we took the core math concepts from the standards listed under each of these categories and rewrote them in short, student-friendly phrases to be displayed on the Math GNOMe board (page 18). We also created a document that teachers could easily reference to guide instruction (page 51). We decided this information should be something that the children look at as well to see what they are expected to learn and know by the end of the school year. To further increase student accountability and awareness of the standards, we created *I Can...* Statements reference charts (pages 19–22). These charts connect the core math concepts on the bulletin board phrases with more specific details of what the standards mean and how students are expected to demonstrate understanding. See Chapter 2 for more information on using these charts when conducting conferences with students.

Lemov (2010) discusses the importance of introducing your students to the standards that your lessons are addressing. Student achievement grows when students are aware of the high expectations placed on them.

GNOMe stands for geometry, number sense, operations, and measurement.

To help students become familiar with the standards, display the Math GNOMe on a bulletin board in your classroom. You want the Math GNOMe to be readily available to you, your students, and anyone who walks into your classroom. This is a great motivator for students and helps to keep you on track with your math instruction (see Chapter 3 for bulletin board setup).

10   The Math GNOMe and Common Core 4 • Gr. 4

Implementing the Math GNOMe

Introducing a New Standard

The Math GNOMe instruction block is set up to be a series of mini-lessons followed by independent practice. The mini-lessons in math instruction revolve around the Math GNOMe. Choose a standard that you would like to teach to your students. Write on a sentence strip or item of your choice the short, student-friendly version of the standard from the list of Math GNOMe bulletin board phrases (page 18). Then add the phrase to the bulletin board under the correct GNOMe category. Proceed with your lesson on that standard, oftentimes bringing your students' attention to the bulletin board phrase. The more you address the Math GNOMe bulletin board and standard you are teaching, the more likely students are to make the connection between the lesson they are engaged in and the standard they are expected to master.

It is important to teach a class mini-lesson when first introducing a new standard to place on the GNOMe bulletin board (see sample mini-lessons on pages 14–17).

> The more you address the Math GNOMe bulletin board and standard you are teaching, the more likely students are to make the connection between the lesson they are engaged in and the standard they are expected to master.

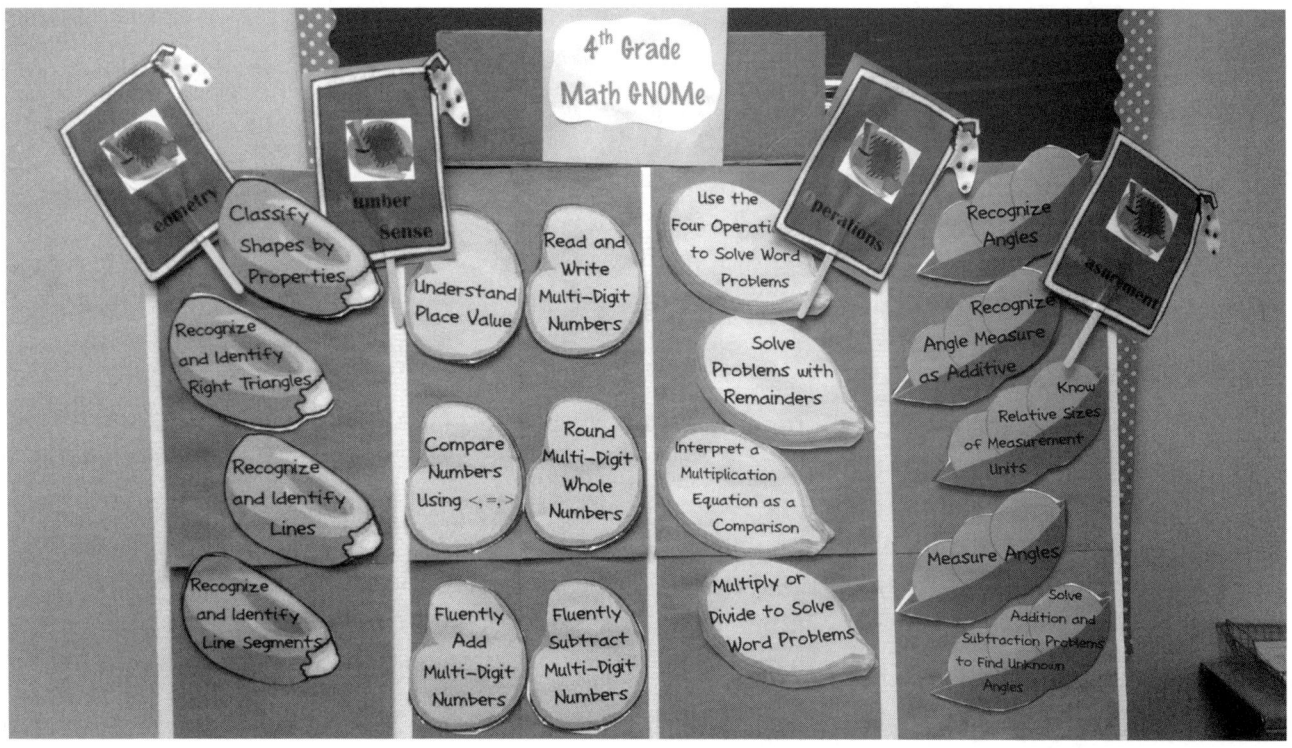

This whole-group setting allows all students to become familiar with the new standard that the class will be working on. It also allows the entire class to revel in the novelty of a new phrase being added to the ever-growing GNOMe bulletin board. After the original lesson, follow-up lessons are a variety of whole-group, small-group, and individual lessons.

Typically, the whole-group lessons occur in the days following the first introduction of the new standard. After the majority of the class has mastered the standard, move to a new standard but continue to work with the remaining students who have not mastered it yet. Finally, focus on individual students who need the one-on-one attention to master the particular skill.

Once the phrase is placed on the Math GNOMe bulletin board and the lesson is taught, it stays there for the remainder of the year. Make a point to constantly refer back to the phrase whenever students are engaged in an activity or discussion that pertains to that specific standard. Strive to teach various mini-lessons concerning that standard throughout the entire school year.

Recording Mastery

During individual student conferences, discuss the standards that the student has not mastered yet. Help the student select a standard to work on. This standard will be the focus of the student's work during Common Core 4 as well as the subject of your future conferences until mastery is

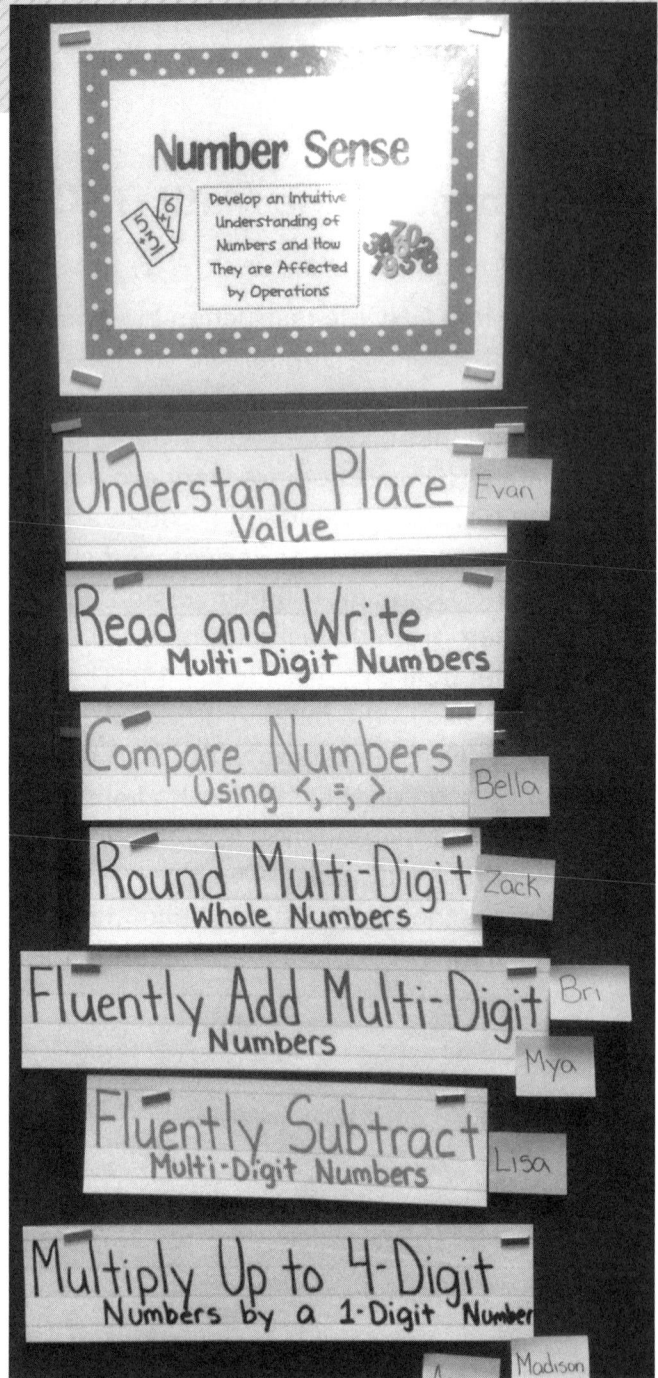

achieved (see Chapter 4 for details). After the conference, the student writes his or her name or assigned class number on a sticky note, an item of your choosing, or a leaf from the companion Math GNOMe décor (see the inside back cover for more details) and attaches it next to the standard on the GNOMe bulletin board. As students master one standard, they should move their name next to another standard on which they need to focus. This encourages students to actively use the Math GNOMe bulletin board and recognize the connection between their learning and the standards. It allows the student to take control of his or her learning and focus on becoming a better mathematician in that domain. It allows you to focus on one-on-one activities that help the student master that specific standard. Keep in mind that according to the Common Core Standards, mastery means the child acquires a deep understanding of the content and the ability to apply that knowledge in a performance task or real-life setting.

The Math GNOMe is meant to be a quick, easy reference tool for both you and your students. It should be used in a way that drives your instruction and your students' learning. There is no set sequence for introducing the domains of the Math GNOMe. For example, you can teach a measurement lesson followed by an operations lesson. You do not need to look at the categories as units. Rather, introduce and revisit the individual standards throughout the entire school year.

Make it a priority to constantly refer back to the Math GNOMe bulletin board, connect your mini-lessons to the standards being displayed, and make sure your students recognize the connection between their learning and the standards they are expected to master.

> As students master one standard, they should move their name next to another standard on which they need to focus.

> The Math GNOMe is meant to be a quick, easy reference tool for both you and your students.

Fourth Grade Geometry
SAMPLE MINI-LESSON

Background

Standard 4.G.A.2 Classify two-dimensional figures based on the presence or absence of parallel or perpendicular lines, or the presence or absence of angles of a specified size. Recognize right triangles as a category, and identify right triangles.

Materials

✓ CHART PAPER OR INTERACTIVE WHITEBOARD
✓ GEOMETRIC SHAPES

Mini-Lesson

1. Show students the Math GNOMe bulletin board phrase "Classify shapes by properties."

2. Review the meaning of "classify" and the names of the two-dimensional shapes that they have learned about previously.

3. Tell students that they will be playing a game called Guess My Rule.

4. Using chart paper, an interactive whiteboard, or actual shapes, show students two different groups of shapes based on parallel or perpendicular lines and/or the size of angles.

5. Tell students that you placed these shapes into two groups based on one rule, and it is their job to figure out what the rule is (e.g., one group has shapes with right angles and the other group has shapes without right angles).

6. Play this game several times. In subsequent days, select students to create the groups and have the rest of the students guess the rules.

7. Direct students' attention to the Math GNOMe bulletin board phrase "Classify shapes by properties." Review the different properties that were used to classify the shapes.

Extension

When you are confident that students can play Guess My Rule independently, add the materials needed to play this game to the math games binder.

Fourth Grade Number Sense
SAMPLE MINI-LESSON

Background

Standard 4.NBT.B.5 Multiply a whole number of up to four digits by a one-digit whole number, and multiply two two-digit numbers, using strategies based on place value and the properties of operations. Illustrate and explain the calculation by using equations, rectangular arrays, and/or area models.

This standard should be broken apart and addressed in several different lessons. The following mini-lesson focuses on the portion of the standard that uses the place value strategy to solve multiplication problems with up to a four-digit whole number multiplied by a one-digit whole number.

Materials

✓ WIPE-OFF MARKERS
✓ INDIVIDUAL WHITEBOARDS (ONE FOR EACH STUDENT)
✓ INTERACTIVE WHITEBOARD, CHART PAPER, WHITEBOARD, OR CHALKBOARD

Mini-Lesson

1 Show students the Math GNOMe bulletin board phrase "Multiply up to a 4-digit number by a 1-digit number." Explain that they will learn a new strategy for multiplying.

2 Present the following word problem: "There are 143 students going on a field trip. The cost of the field trip is $8.00 per person. How much money will the class need so that all of the students can attend the field trip?"

3 Ask students how they can solve the problem. They should determine that multiplication is the most efficient way to find a solution.

4 Ask students to write the number 143 in expanded form, i.e., break it down into hundreds, tens, and ones (100 + 40 + 3). Have them visualize what each of those numbers would look like in an area if each number was a length multiplied by a width of 8. Draw the following area model:

143 =	100	+	40	+	3
8	8 x 100 = 8 x 1 HUNDREDS = 8 HUNDREDS		8 x 40 = 8 X 4 TENS = 32 TENS		8 x 3 = 24

8 x 143 = 800 + 320 + 24 = 1,144

5 Demonstrate the area model for multiplication. If students do not understand, use smaller numbers in the area model (for example, 10 x 8).

6 Once you are confident that students understand the strategy, have them try several problems using small whiteboards or pieces of paper.

7 Direct students' attention to the Math GNOMe bulletin board phrase "Multiply up to a 4-digit number by a 1-digit number." Review the area model for solving multiplication problems.

Extension

Students can use this strategy to multiply a two-digit number by a two-digit number. This is a visual model that allows visual learners to see multiplication happening.

Fourth Grade Operations
SAMPLE MINI-LESSON

Background

Standard 4.OA.A.1 Interpret a multiplication equation as a comparison, e.g., interpret 35 = 5 × 7 as a statement that 35 is 5 times as many as 7 and 7 times as many as 5. Represent verbal statements of multiplicative comparisons as multiplication equations.

This lesson introduces the tape diagram and the Read, Draw, Write (RDW) process. Continue to use this model with other multiplicative comparison word problems that tend to be confusing to students.

Materials
✓ INTERACTIVE WHITEBOARD, WHITEBOARD, OR CHART PAPER

Mini-Lesson

1 Show students the Math GNOMe bulletin board phrase "Interpret multiplication equations as comparisons." Explain that they will learn how to represent a comparison word problem to help them determine which operation is needed to solve it.

2 Write the following problem for students to see: *An adult tiger will eat three times a much food as a baby tiger. The adult tiger eats 30 pounds of food. How many pounds of food do an adult and baby tiger eat altogether?*

3 Explain to the students that they will be using a new math tool called a tape diagram to help them solve problems like this. Students will also be using a word problem model called Read, Draw, Write to help them make sense of problems and persevere in solving them, which are objectives of the Common Core Standards.

4 To use the RDW process, read the first sentence. Then stop and ask: *Can I draw something? Can I label something? Can I write something? What do I see? What do I notice?*

5 Continue this process of drawing, labeling, and writing until the students are able to visualize the problem. Then depict the problem by drawing the following tape diagram:

ADULT TIGER [| |] } 30 lbs } ? lbs
BABY TIGER [   ] } ? lbs

A = NUMBER OF POUNDS THE ADULT TIGER EATS
B = NUMBER OF POUNDS THE BABY TIGER EATS
3 × B = A
3 × B = 30 POUNDS
B = 10 POUNDS
A + B = 30 + 10
A + B = 40 POUNDS OF FOOD

6 Direct students' attention to the Math GNOMe bulletin board phrase "Interpret multiplication equations as comparisons." Review the RDW process and tape model strategies.

Extension

Continue this activity over several days and weeks until the standard is mastered. Students may also use the RDW process and tape diagrams when solving other word problems to help students "make sense of problems and persevere in solving them." As students use RDW and tape diagrams, they will begin to see how their drawings help them make sense of the problem.

Fourth Grade Measurement
SAMPLE MINI-LESSON

Background

Standard 4.MD.A.2 Use the four operations to solve word problems involving distances, intervals of time, liquid volumes, masses of objects, and money, including problems involving simple fractions or decimals, and problems that require expressing measurements given in a larger unit in terms of a smaller unit. Represent measurement quantities using diagrams such as number line diagrams that feature a measurement scale.

Materials

- ✓ WORD PROBLEMS INVOLVING MEASUREMENT (E.G., DISTANCE, TIME, VOLUME, MASS, AND MONEY)
- ✓ SMALL INDIVIDUAL WHITEBOARDS
- ✓ MARKERS
- ✓ WHITEBOARD ERASERS

Mini-Lesson

1. Show students the Math GNOMe bulletin board phrase "Solve word problems involving measurement."

2. Tell the students that they will be working on a challenge today. Remind them of the strategies they know of to solve word problems in which they need to add, subtract, multiply, and divide (e.g., number bonds, tape diagrams, pictures). Students will now apply these strategies to solve word problems involving measurement.

3. Write the following problem for students to see: *What time does Marla have to leave to be at her friend's house by a quarter after three if the trip takes 90 minutes?*

4. Give students a chance to solve the problem on their individual whiteboards. Ask students to turn their boards over when they are done.

5. Have students share their strategies with a partner. Ask two students who used different strategies to explain how they reached their answers. Then have students clear their whiteboards and get ready for the next problem.

6. Write the following problem for students to see: *Stephen wants to make fruit punch using two flavors of juice. There will be a total of 320 ml of juice in his mug. He will add 80 ml of pineapple juice to his mug and three times as much orange juice as pineapple. How much orange juice will he add?*

7. Have students solve the problem on their whiteboards and share their strategies with a partner. Then have two students who used different strategies explain how they reached their answers.

8. Finish the mini-lesson by reviewing the Math GNOMe bulletin board phrase "Solve word problems involving measurement" and how it relates to the day's mini-lesson.

The Math GNOMe
FOURTH GRADE BULLETIN BOARD PHRASES

Geometry

Recognize and identify lines *4.G.A.1*

Recognize and identify line segments *4.G.A.1*

Recognize and identify rays *4.G.A.1*

Recognize and identify angles *4.G.A.1*

Recognize and identify perpendicular lines *4.G.A.1*

Recognize and identify parallel lines *4.G.A.1*

Classify shapes by properties *4.G.A.2*

Recognize and identify right triangles *4.G.A.2*

Recognize and identify symmetrical shapes *4.G.A.3*

Number Sense
(NUMBER & OPERATIONS IN BASE TEN AND NUMBER & OPERATIONS—FRACTIONS)

Understand place value *4.NBT.A.1*

Read and write multi-digit whole numbers *4.NBT.A.2*

Compare numbers using <, =, > *4.NBT.A.2*

Round multi-digit whole numbers *4.NBT.A.3*

Fluently add multi-digit numbers *4.NBT.B.4*

Fluently subtract multi-digit numbers *4.NBT.B.4*

Multiply up to a 4-digit number by a 1-digit number *4.NBT.3.5*

Multiply two 2-digit numbers *4.NBT.B.5*

Find whole number quotients and remainders *4.NBT.B.6*

Understand equivalent fractions *4.NF.A.1, 4.NF.C.5*

Compare fractions with <, =, > *4.NF.A.2*

Understand what a fraction is *4.NF.B.3*

Add and subtract fractions *4.NF.B.3a, 4.NF.C.5*

Break apart a fraction into a sum of fractions *4.NF.B.3b*

Add and subtract mixed numbers *4.NF.B.3c*

Solve word problems involving fractions *4.NF.B.3d, 4.NF.B.4c*

Multiply fractions by whole numbers *4.NF.B.4, 4.NF.B.4a, 4.NF.B.4b*

Understand decimal notation for fractions *4.NF.C.6*

Compare decimals with <, =, > *4.NF.C.7*

Operations
(OPERATIONS & ALGEBRAIC THINKING)

Interpret multiplication equations as comparisons *4.OA.A.1*

Multiply or divide to solve word problems *4.OA.A.2*

Use the four operations (+, −, ×, ÷) to solve word problems *4.OA.A.3*

Solve problems with remainders *4.OA.A.3*

Use a letter to stand for an unknown quantity *4.OA.A.3*

Recognize that a whole number is a multiple of its factors *4.OA.B.4*

Find factor pairs *4.OA.B.4*

Identify prime or composite numbers *4.OA.B.4*

Determine if a whole number is a multiple of a given number *4.OA.B.4*

Create a number or shape pattern *4.OA.C.5*

Identify features of patterns *4.OA.C.5*

Measurement
(MEASUREMENT & DATA)

Know relative sizes of measurement units *4.MD.A.1*

Record measurement equivalents in two-column tables *4.MD.A.1*

Solve word problems involving measurement *4.MD.A.2*

Represent measurement quantities using diagrams *4.MD.A.2*

Apply the perimeter formula *4.MD.A.3*

Apply the area formula *4.MD.A.3*

Create a line plot *4.MD.B.4*

Use and interpret data in line plots *4.MD.B.4*

Recognize angles *4.MD.C.5*

Measure angles *4.MD.C.5a, 4.MD.C.5b, 4.MD.C.6*

Read angle degrees *4.MD.C.5a, 4.MD.C.5b*

Understand how to use a protractor *4.MD.C.6*

Recognize angle measure as additive *4.MD.C.7*

Solve addition and subtraction problems to find unknown angles *4.MD.C.7*

Geometry I Can... Statements

STANDARD NUMBER	BULLETIN BOARD PHRASE(S)	I CAN STATEMENT(S)
4.G.A.1	Recognize and identify lines	I can draw points, lines, and line segments.
	Recognize and identify line segments	I can identify lines and line segments in 2-D figures.
	Recognize and identify rays	I can draw rays and angles (right, acute, obtuse).
	Recognize and identify angles	I can identify rays and angles in 2-D figures.
	Recognize and identify perpendicular lines	I can identify perpendicular lines in 2-D figures. I can draw perpendicular lines.
	Recognize and identify parallel lines	I can identify parallel lines in 2-D figures. I can draw parallel lines.
4.G.A.2	Classify shapes by properties	I can classify 2-D figures based their lines and angles.
	Recognize and identify right triangles	I can recognize and identify right triangles within 2-D figures.
4.G.A.3	Recognize and identify symmetrical shapes	I can identify figures that can be divided symmetrically.
		I can recognize and draw lines of symmetry in 2-D figures.

Number Sense *I Can...* Statements

STANDARD NUMBER	BULLETIN BOARD PHRASE(S)	I CAN STATEMENT(S)
4.NBT.A.1	Understand place value	I can determine that a digit represents ten times what it would be in the place to its right.
4.NBT.A.2	Read and write multi-digit whole numbers	I can read and write multi-digit whole numbers. I can read and write number names. I can read and write multi-digit numbers in expanded form.
	Compare numbers using <, =, >	I can compare multi-digit numbers using <, =, or >.
4.NBT.A.3	Round multi-digit whole numbers	I can round multi-digit whole numbers to any place.
4.NBT.B.4	Fluently add multi-digit numbers Fluently subtract multi-digit numbers	I can fluently add multi-digit numbers. I can fluently subtract multi-digit numbers.
4.NBT.B.5	Multiply up to a 4-digit number by a 1-digit number	I can multiply up to a four-digit number by a one-digit number. I can illustrate and explain the calculation using equations, rectangular arrays, and/or area models.
	Multiply two 2-digit numbers	I can multiply two two-digit numbers. I can illustrate and explain the calculation using equations, rectangular arrays, and/or area models.
4.NBT.B.6	Find whole-number quotients and remainders	I can find quotients and remainders with up to four-digit dividends and one-digit divisors. I can illustrate and explain the calculation.
4.NF.A.1, 4.NF.C.5	Understand equivalent fractions	I can use visual fraction models to recognize equivalent fractions. I can use visual fraction models to explain equivalent fractions. I can use visual models to generate equivalent fractions. I can express a fraction with denominator 10 as an equivalent fraction with denominator 100, and use this technique to add two fractions.
4.NF.A.2	Compare fractions with <, =, >	I can use <, =, or > to compare two fractions with different numerators and denominators. I can justify the calculation with a fraction model.
4.NF.B.3	Understand what a fraction is	I can understand a fraction is a/b with a>1 as a sum of fractions 1/b.
4.NF.B.3a, 4.NF.C.5	Add and subtract fractions	I can understand that addition and subtraction of fractions is joining and separating parts of the same whole.
4.NF.B.3b	Break apart a fraction into a sum of fractions	I can break apart a fraction into a sum of fractions with the same denominator in more than one way.
4.NF.B.3c	Add and subtract mixed numbers	I can add and subtract mixed numbers with like denominators.
4.NF.B.3d, 4.NF.B.4c	Solve word problems involving fractions	I can solve addition, subtraction, multiplication, and division word problems about fractions referring to the same whole and having like denominators.
4.NF.B.4, 4.NF.B.4a, 4.NF.B.4b	Multiply fractions by whole numbers	I can multiply a fraction by a whole number.
4.NF.C.6	Understand decimal notation for fractions	I can use decimal notation for fractions with denominators 10 or 100.
4.NF.C.7	Compare decimals with <, =, >	I can use <, =, or > to compare two decimals to hundredths. I can justify the conclusions.

Operations *I Can...* Statements

STANDARD NUMBER	BULLETIN BOARD PHRASE(S)	I CAN STATEMENT(S)
4.OA.A.1	Interpret multiplication equations as comparisons	I can explain how a multiplication equation can be used to compare.
4.OA.A.2	Multiply or divide to solve word problems	I can multiply or divide to solve word problems.
4.OA.A.3	Use the four operations (+, −, ×, ÷) to solve word problems	I can use +, −, ×, ÷ to solve multistep word problems with whole-number answers.
	Solve problems with remainders	I can use +, −, ×, ÷ to solve multistep word problems with whole-number answers and remainders.
	Use a letter to stand for an unknown quantity	I can write equations with letters standing for the unknown quantities.
4.OA.B.4	Determine if a whole number is a multiple of a given number	I can determine if a whole number is a multiple of a given number.
	Recognize that a whole number is a multiple of its factors	I can recognize that a whole number is a multiple of each of its factors.
	Find factor pairs	I can find factor pairs for whole numbers 1–100.
	Identify prime or composite numbers	I can determine whether a whole number (1–100) is prime or composite.
4.OA.C.5	Create a number or shape pattern	I can generate a number or shape pattern that follows a given rule.
	Identify features of patterns	I can identify features of the pattern.

Measurement *I Can...* Statements

STANDARD NUMBER	BULLETIN BOARD PHRASE(S)	I CAN STATEMENT(S)
4.MD.A.1	Know relative sizes of measurement units	I know relative sizes of measurement units within a system of units.
	Record measurement equivalents in two-column tables	I can express measurements in a larger unit in terms of a smaller unit.
		I can record measurement equivalents in a two-column table.
4.MD.A.2	Solve word problems involving measurement	I can use +, −, ×, ÷ to solve word problems involving different units of measurement.
	Represent measurement quantities using diagrams	I can represent measurement quantities using diagrams.
4.MD.A.3	Apply the perimeter formula	I can apply the perimeter formula for rectangles to real-world and mathematical problems.
	Apply the area formula	I can apply the area formula for rectangles to real-world and mathematical problems.
4.MD.B.4	Create a line plot	I can make a line plot using fractions of a unit.
	Use and interpret data in line plots	I can use information in line plots to solve problems involving adding and subtracting fractions.
4.MD.C.5	Recognize angles	I can recognize that angles are geometric shapes formed when two rays share an endpoint.
4.MD.C.5a, 4.MD.C.5b	Read angle degrees	I can read the degree of an angle.
4.MD.C.5a, 4.MD.C.5b, 4.MD.C.6	Measure angles	I can understand how an angle in a circle is measured.
4.MD.C.6	Understand how to use a protractor	I can use a protractor to construct and measure angles.
4.MD.C.7	Recognize angle measure as additive	I can recognize that the sum of the angle parts is equal to the whole angle.
	Solve addition and subtraction problems to find unknown angles	I can solve addition and subtraction problems with unknown angles on a diagram.

CHAPTER 2

The Common Core 4

ABOUT THE COMMON CORE 4

The Common Core 4 refers to the four independent practice activities that students participate in on a consistent basis within the math block.

As stated earlier, the math block is set up as a series of mini-lessons followed by independent practice. Students select the Common Core 4 activity they will work on and the teacher records the selection on the Common Core 4 Student Choice Log (page 49). This independent practice time allows teachers to conduct short conferences with individual and small groups of students. This section describes the four parts that make up the Common Core 4 and explains how they relate to the standards and what you can do to ensure successful implementation in your classroom.

While the Math GNOMe refers to the content that you are teaching, the Common Core 4 is the mathematical practice students are working on independently.

The Common Core 4 activities that students engage in address the Common Core's Mathematical Practice Standards. These standards require teachers to offer students the opportunity to "make sense of problems and persevere in solving them, reason abstractly and quantitatively, construct viable arguments and critique the reasoning of others, model with mathematics, use appropriate tools strategically, attend to precision, look for and make use of structure, and look for and express regularity of repeated reasoning." These standards are an inherent part of the Common Core 4.

Implementing the Common Core 4

Taking the time to discretely teach your expectations, model strict routines, and establish them in your classroom is key to successful implementation of the Common Core 4. It takes several weeks from the first time an aspect of the Common Core 4 is introduced until it is up and running in the way it will be throughout the rest of the school year. In Lemov's (2010) book *Teach Like a Champion*, he discusses the importance of establishing routines during the first several weeks of school. You are setting the stage for the rest of your school year and gaining more instructional time for the latter school months.

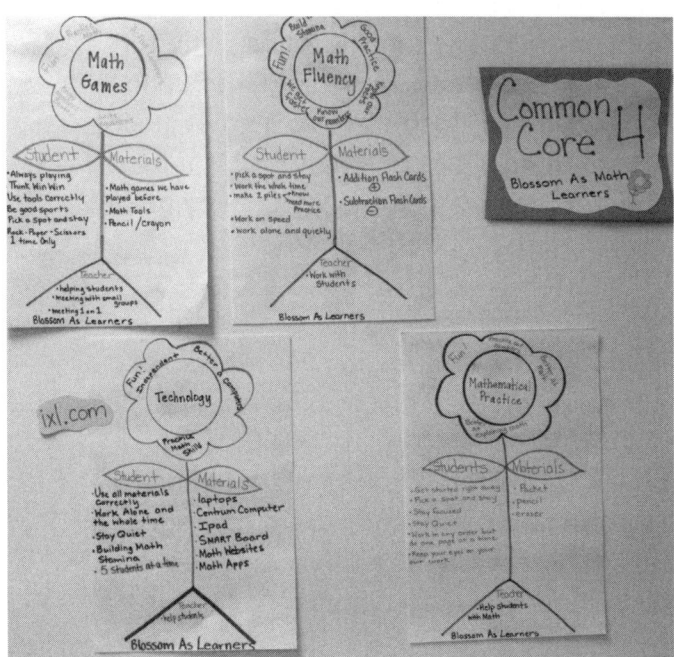

Introduce one of the following Common Core 4 components at a time, allowing students to work toward mastering that activity before the next one is introduced. This means that the Common Core 4 will not be fully up and running in your classroom for several weeks. If you take this time to set up the routines, your students will work independently for the rest of the year. This, in turn, allows for uninterrupted time for instruction and assessment.

Math Fluency

The Common Core Standards emphasize numeracy fluency. Rapid math-fact retrieval has been shown to be a strong predictor of performance on mathematics achievement tests. Russell (2000) discusses the importance of computational fluency and its effects on students' mathematical achievement. A student who can quickly compute the answers to basic facts is able to devote more of his or her thinking capacity to higher-level concepts and problem solving. Therefore, students should spend time every day practicing their math facts to increase their automaticity.

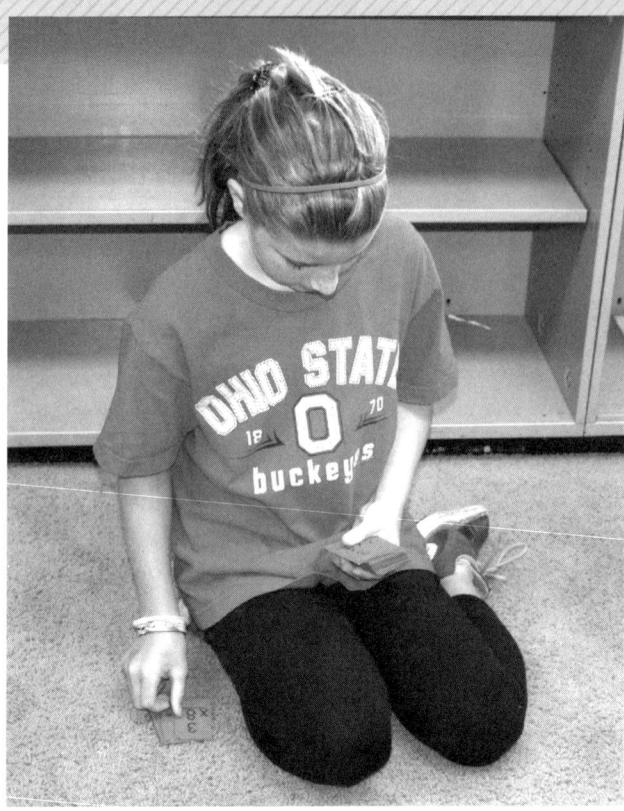

> **Rapid math-fact retrieval has been shown to be a strong predictor of performance on mathematics achievement tests.**

The Common Core 4 provides students with the time to work on math fluency. It is important to view this time as an opportunity for shared learning as discussed in the Common Core Standards. Each student should have one or more sets of flash cards that can be used during independent work or with partners and small groups. When students choose to work on math fluency, they take their flash cards to a quiet place to drill themselves or a partner. Groups of students can play numeracy fluency games involving the math facts cards.

When first introducing the concept of math fluency to students, describe what it means and then discuss the importance of knowing math facts. Have students brainstorm different ways math fluency is important to them, and record their ideas on a Blossom Chart. A Blossom Chart is where you will record students' thoughts

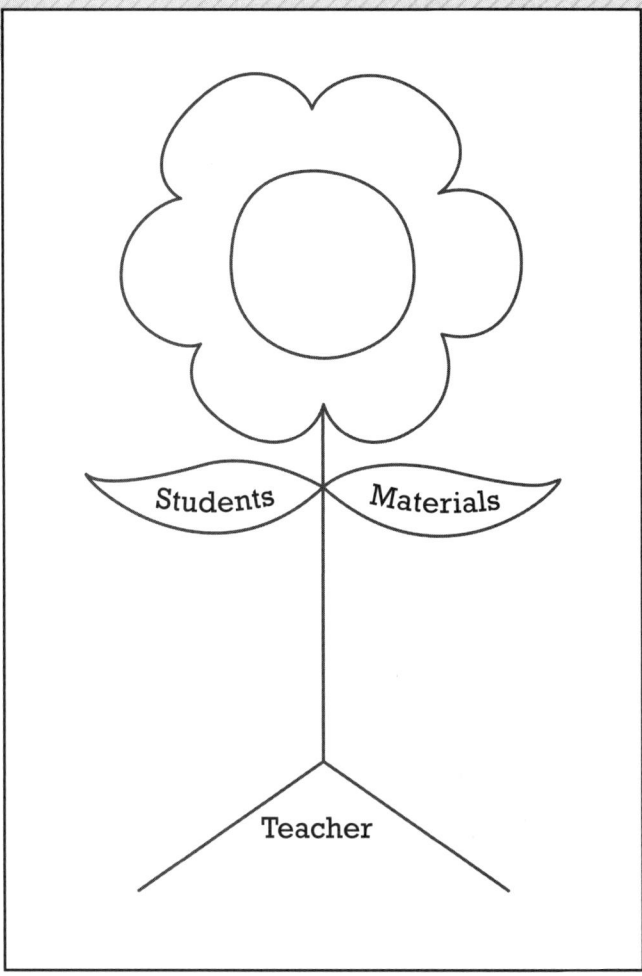

and ideas about the Common Core 4. This reference chart outlines materials and student and teacher expectations that create an environment that is conducive to learning during each Common Core 4 activity. When your students' needs are met, they are able to blossom as learners. After introducing the Math Fluency Blossom Chart, ask students to come up with ways they can practice math fluency. Guide the discussion so students recognize that the only way to improve their math fluency is to practice.

Next, give students time to explore the math facts cards. Allowing students time to investigate the materials before the lesson minimizes the distraction of new materials. After they have explored the cards, ask students to help finish the Blossom Chart for Math Fluency. First, under the materials section, write "math facts cards." Next, have students come up with ideas for what the classroom should look like and sound like and what they should be doing during this time. Record these suggestions in the student expectations section of the Blossom Chart. Guide the discussion and make sure the Blossom Chart includes phrases such as *get started right away, work quietly, make smart choices about where to sit,* and *work the entire time.*

This reference chart outlines materials and student and teacher expectations that create an environment that is conducive to learning.

Ask students questions such as "How should you organize your cards as you go through them?" "What should you do if you finish all of your math cards?" You might suggest that students create two piles of cards while they work—one pile for the facts that they know and can recall quickly and one pile for the facts that they do not know or cannot recall quickly. Then students can go back and focus on the facts with which they need the most practice.

For the final step of the Blossom Chart, have students brainstorm what the teacher should be doing during this time. Oftentimes students suggest that the teacher should be helping students learn math, or something along those lines.

Once the Blossom Chart is finalized, have several students model what a round of Math Fluency time should look like. Invite the rest of the class to observe the behaviors of students during this modeling session. Critique the student volunteers out loud so the class understands what students are doing correctly. Then ask the class to describe what they saw and heard and compare it to the Blossom Chart.

End the first lesson by giving students an opportunity to practice. Because this will be students' first encounter with Math Fluency time in the Common Core 4 setting, have students practice for only 1–2 minutes at a time. Discuss areas in the classroom where students can work, such as at a desk, on the floor, against a pillow, or on the rug. Call students one at a time to choose a place to work in the classroom. Once all of the students are situated around the classroom, start your timer. Monitor students to make sure they are following the expectations outlined on the Blossom Chart. However, try to avoid communicating with students in any way. Students need to get used to the idea that this is an independent activity. Once the Common Core 4 is fully up and running,

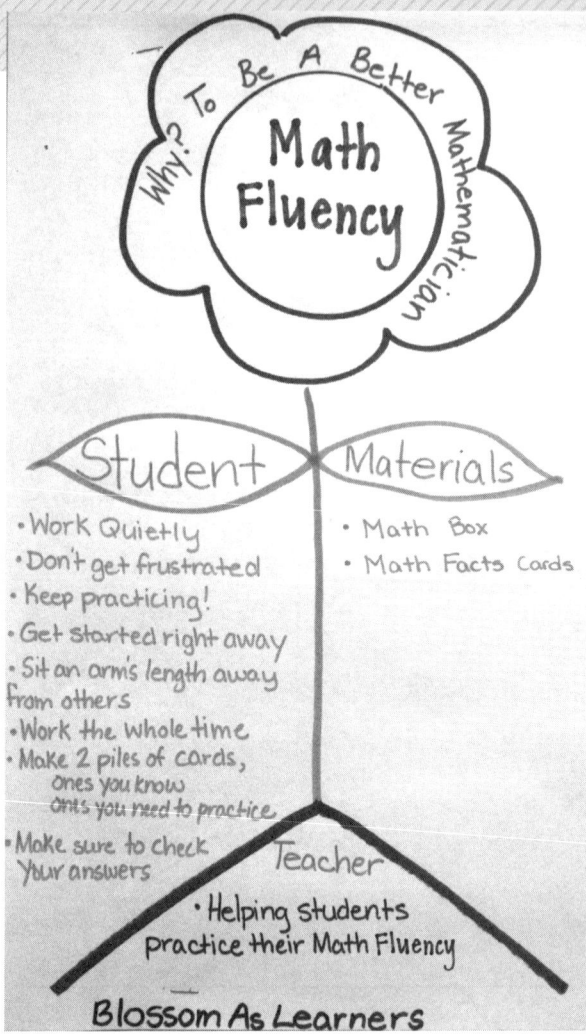

the teacher will be meeting with students and groups and should not be interrupted.

Next, signal to students when time is up. The signal can be a flash of the lights, ringing a bell, clapping your hands, and so on, and should be consistent for all Common Core 4 activities. The expectation is that students should quickly put away materials and quietly return to the classroom meeting area. You may need to practice picking up and moving to your meeting area several times. Once students have gathered together again, have another class discussion about how they think they did with the first round of the Math Fluency activity. Review the Blossom Chart, and make any changes or additions as necessary. If time permits, have students practice one or two more 1- to 2-minute practice rounds of Math Fluency time on the first day.

> **It is important to remember that before more practice time is added, students must be meeting the expectations stated on the Blossom Chart.**

The second day of Math Fluency practice looks very similar to the first day. Invite students to gather together at your meeting area. Review the Blossom Chart and what Math Fluency time should look and sound like. Ask for student volunteers to model a round of Math Fluency practice while the rest of the class observes and critiques out loud. Then have the whole class practice the Math Fluency activity. Increase the total time to 2–3 minutes.

After this time is up, signal to the students to pick up their cards quickly and quietly and come to the meeting area. Discuss with the class what went well during the round, what needs to be worked on, and anything that needs to be changed or added to the Blossom Chart. Then have students practice for two or three more 2- to 3-minute rounds of Math Fluency time that day.

For subsequent days, continue to have students practice math fluency and their math facts and have class discussions about what is going well and what needs to be worked on. Each day, add 1 or 2 minutes to the rounds until you eventually reach your goal—the amount of time each round will be for the rest of the year. This goal should be anywhere from 5 to 15 minutes, depending on your students and the amount of time you have available for math instruction.

It is important to remember that before more practice time is added, students must be meeting the expectations stated on the Blossom Chart. If one or more students are struggling to stay on task, you may need to reduce the time and work toward mastery within that given time for several days. It is OK to not add more minutes each day. It is more important that students are displaying mastery within the given time frame, even if it means taking longer than expected to reach your final goal.

Weekly Plans

MATH FLUENCY

Day 1

1. Whole-class discussion on importance of math fluency
2. Brainstorm how to study in order to attain math fluency
3. Explore Math Fluency practice materials
4. Create Math Fluency Blossom Chart
5. Students model Math Fluency activity
6. Practice round for 1–2 minutes
7. Review and edit Blossom Chart
8. One or two more practice rounds for 1–2 minutes each

Day 2

1. Review Math Fluency Blossom Chart
2. Students model Math Fluency activity
3. Practice round for 2–3 minutes
4. Review and edit Blossom Chart
5. One or two more practice rounds for 2–3 minutes each

Day 3

1. Review Math Fluency Blossom Chart
2. Students model Math Fluency activity
3. Practice round for 3–4 minutes
4. Review and edit Blossom Chart
5. One or two more practice rounds for 3–4 minutes each

Day 4–10

1. Review Math Fluency Blossom Chart
2. Practice round: Add 1–2 minutes each day
3. Review and edit Blossom Chart as needed
4. One or two more practice rounds if time permits

Day 11–Beyond

1. Review Math Fluency Blossom Chart as needed
2. One full round of Math Fluency practice for 5 to 15 minutes
3. Review and edit Blossom Chart as needed

Day 1 Lesson Plan

MATH FLUENCY

Objectives

▸ Introduce Math Fluency time to students.

▸ Allow students to familiarize themselves with materials.

▸ Create Math Fluency Blossom Chart.

▸ Have students participate in several practice rounds.

Materials

✓ MATH FACTS CARDS

Lesson Outline

CLASS DISCUSSION

▸ What is math fluency?

▸ Why is math fluency important?

▸ What can math fluency help us do?

▸ Let's record our ideas on the Math Fluency Blossom Chart.

POSSIBLE STUDENT RESPONSES

▸ Numbers you put together.

▸ Numbers you know from memory.

▸ Math fluency helps us do math.

▸ Math fluency helps us be better at math.

THE EXPECTATIONS

▸ The goal in fourth grade is for every student to fluently add and subtract multi-digit numbers. What things can we do to reach this goal?

▸ The only way to get better at your fluency is to spend time practicing. We will have time each day for you to practice.

POSSIBLE STUDENT RESPONSES

▸ Practice math fluency.

▸ Talk about math fluency.

▸ Talk about math facts.

THE BLOSSOM CHART

▸ Let's fill out our Blossom Chart.

▸ What should we put in the "Materials" section of our Blossom Chart?

▸ What are the student expectations during Math Fluency time? What should students be doing? What should our classroom look like? What should our classroom sound like?

▸ How should you organize your cards?

▸ What should you do if you finish your pile of cards before time is up?

▸ What are the teacher expectations during Math Fluency time? What should the teacher be doing?

POSSIBLE STUDENT RESPONSES

▸ Math facts cards and math boxes go in the "Materials" section.

▸ Students should be focused and work the whole time.

▸ Students should work quietly.

▸ Students should sit far away from friends.

▸ Make piles of cards you know and cards you don't know.

▸ If you finish before time is up, you should start over.

▸ The teacher is meeting with students.

The Math GNOMe and Common Core 4 • Gr. 4

MODEL

- Ask a student volunteer to show the class what Math Fluency practice should look like.
  - Critique the student as he or she models the Math Fluency activity.
    - *I see two piles of cards being made.*
    - *I see (student name) working quietly.*
    - *I see (student name) sitting far away from friends.*
    - *I see (student name) staying focused and working the whole time.*
- *What did you see (student name) doing right?*
- *Did (student name) follow all of the expectations on our Blossom Chart?*

POSSIBLE STUDENT RESPONSES

- I saw (student name) working the whole time.
- I saw (student name) working quietly.
- I think (student name) could have stayed focused a little more.
- I like how (student name) made two neat piles of cards.

PRACTICE

- Invite students to practice a round of Math Fluency time.
- Call students one at a time to choose a spot in the classroom to work.
- Once all students are working, start timing.
- Stop the round if any student is off task or not following the Blossom Chart expectations.
- Signal to students that the round is over.
- Have students practice quickly picking up their cards and quietly returning to the classroom meeting area.

REVIEW

- *How did that round go?*
- *What items on our Blossom Chart did you see happening?*
- *Is there anything we need to work on?*
- *Is there anything we need to add to the Blossom Chart?*
- Complete another practice round and review if time permits.

POSSIBLE STUDENT RESPONSES

- I think we could have worked a little more quietly.
- I think we did a good job staying focused.
- I think we need to try to come back to the meeting area a little faster.
- We should write on the Blossom Chart that you shouldn't sit next to your friends because they might distract you.

Math Games

Part of the reason we panicked when we heard about the Common Core Standards was because we envisioned having to spend countless hours creating math games that align with the Common Core Standards. However, relief set in once we created the Math GNOMe, which helped us to realize that we did not need to reinvent the wheel. Many of the math games we were already using aligned well with the Common Core Standards. If they did not, there was usually a way we could adjust the game so it was aligned to the standards.

Math games are a great way to motivate students. Students like playing competitive games in the classroom setting. The games help students enjoy the math concept being learned. Through math games, students are engaged in the mathematical practices of reasoning abstractly and quantitatively, using appropriate tools strategically, and modeling what they have learned with mathematics. Games are a great way to get students to practice essential skills that are being taught in the classroom. Petsche (2001) reports on the effects of using games in the classroom to both excite and engage students in their own learning. Because math games are enjoyable for students, they are a great medium for getting students to practice the math skills addressed in the Common Core Standards.

> **Games are a great way to get students to practice essential skills that are being taught in the classroom.**

Introduce Math Games time to your students in the same way you introduced Math Fluency practice. On the first day, gather students together in the common meeting area. Have a whole-class discussion on the relevance of math games in the classroom and the impact they can have on student learning. Brainstorm ways that math games can help students become better mathematicians and record them on the Math Games Blossom Chart. Then inform students that they will learn and play many new math games that will help them practice new skills throughout the school year. Before continuing on with the Blossom Chart and introducing students to the first math game of the year, it is important to set the stage for Math Games time. Because manipulatives play such a major role in most math games, it is a good idea to have a mini-lesson on proper use of manipulatives. Within your lesson you should discuss that manipulatives are tools, not toys.

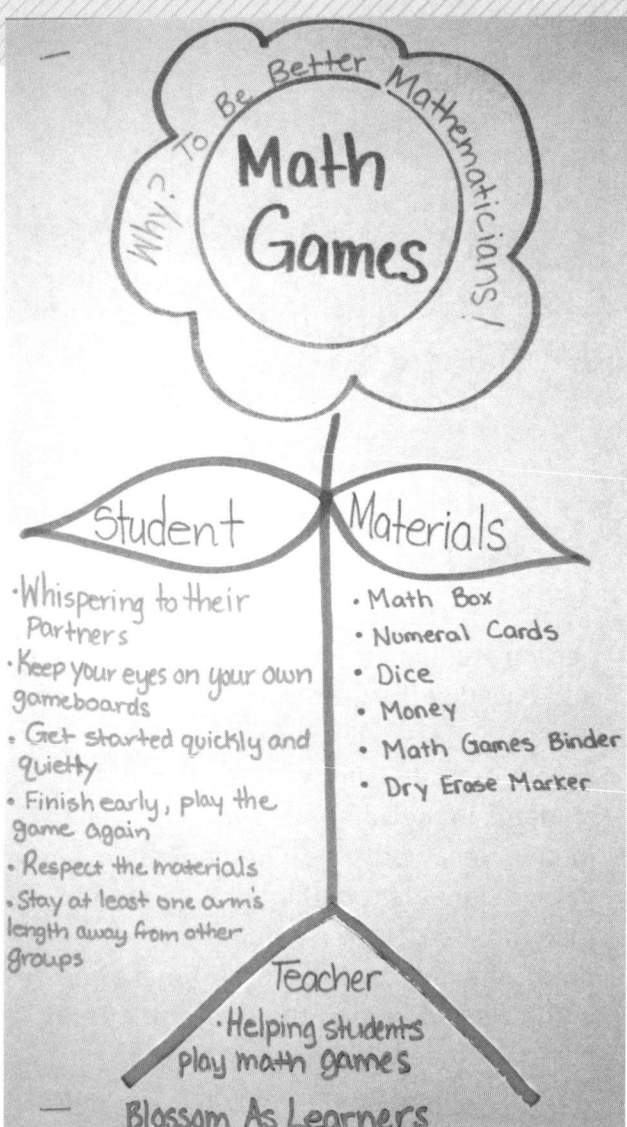

Math Games time. It is important that the Blossom Chart include information about students making smart choices when selecting a partner, starting to play the game right away, playing the game correctly the whole time, and using quiet voices while playing the game. Make sure that students recognize the importance of respecting the materials and picking them up the correct way at the end of the round.

After the Blossom Chart is created, it is time to introduce the first math game of the year. Before discussing what the math game is and the directions for playing, distribute the game materials. As you did with the Math Fluency cards, allow students a few minutes to explore the materials. Once students have had time to see what materials they will be using, bring their attention back to the game they will be learning. Tell students the name of the game and inform them of the Common Core standard that this math game will help them master. Make sure to refer to the Math GNOMe bulletin board.

Tell students that they will need to come up with rules for playing math games and using manipulatives in order for Math Games time to be a successful part of the math block. Work on creating the Blossom Chart for Math Games with the whole class. First, list the different materials students will be using during Math Games time. This list usually includes decks of cards, dice, cubes, and so on.

Then ask students what the classroom should look like and sound like during

Next, explain the game directions to students; then play a sample round of the game with one student volunteer. Be sure to model not only the correct way to play the game according to the game's directions but also the correct behavior that should be exhibited according to the Blossom Chart. After you and the student volunteer model the game for a few minutes, choose two students to model the game and the correct behavior for the whole class. While students model the game, the rest of the class observes and critiques out loud what they see their classmates doing and how their behavior relates to the Blossom Chart expectations.

> **Discuss what went well during the round and what students need to work on.**

During the first practice round, use your preferred method for creating student pairs. Once all students have been matched up with a partner, make sure that student pairs have the materials needed for the game, have selected a spot in the classroom to work, and have gotten started right away. Once all students are playing the game, start with a 4- to 5-minute round. This will ensure that each student has a chance to play his or her part of the game a few times during the practice round.

During this time, monitor student behavior and the noise level. If you notice students are off task or are too noisy, immediately stop the round, have students move back to the meeting area, and discuss what went wrong. Then try again for another 4- to 5-minute round. If you have to stop the round early again, review the Blossom Chart with students and consider reducing the number of minutes for the practice round.

Once they are successful with a round, signal to students to clean up their materials quickly and quietly. The signal you use should be the same signal that you use for Math Fluency activities and all of the Common Core 4 choices. Then invite students back to the classroom meeting area. Discuss what went well during the round and what students need to work on. Revisit the Blossom Chart for Math Games time and change or add anything that is needed. If time permits, have students practice for two or three more 4- to 5-minute rounds.

The second day is very similar to the first day. Begin by reviewing the Blossom Chart for Math Games time and the directions for the game introduced the first day. Have several students model the correct way to play the math game while the other students observe and critique out loud. Then let students play several practice rounds. Add 1 or 2 minutes to the time limit of the previous day depending on what you feel your students can handle.

Be sure that students have mastered one game before introducing another game. Once you have introduced more than one math game and students are still working on becoming independent, choose which game students should play. When you are confident that students are independent in Math Games time, allow them to choose a game that has been previously introduced. Increase the length of each round by 1 or 2 minutes each day until you reach your goal. Your goal should be between 10 and 15 minutes, depending on your students and how much time you have dedicated to your math block.

Weekly Plans

MATH GAMES

Day 1

1. Whole-class discussion on importance of math games
2. Brainstorm how math games help students learn math
3. Explore Math Games materials
4. Create Math Games Blossom Chart
5. Teacher introduces first math game
6. Teacher models the math game with a student
7. Students model the math game
8. Practice round for 4–5 minutes
9. Review and edit Blossom Chart
10. One or two more practice rounds for 4–5 minutes each

Day 2

1. Review Math Games Blossom Chart
2. Students model math game from Day 1
3. Practice round for 5–6 minutes
4. Review and edit Blossom Chart
5. One or two more practice rounds for 5–6 minutes each

Day 3

1. Review Math Games Blossom Chart
2. Students model math game from Day 1
3. Practice round for 6–7 minutes
4. Review and edit Blossom Chart
5. One or two more practice rounds for 6–7 minutes each

Day 4–10

1. Review Math Games Blossom Chart
2. Practice round: Add 1–2 minutes each day
3. Review and edit Blossom Chart as needed
4. One or two more practice rounds if time permits

Day 11–Beyond

1. Review Math Games Blossom Chart as needed
2. One full round of Math Games time for 10 to 15 minutes
3. Review and edit Blossom Chart as needed
4. Teach and model new math games throughout the school year

Day 1 Lesson Plan

MATH GAMES

Objectives

▸ Introduce Math Games time to students.

▸ Allow students to familiarize themselves with materials.

▸ Create Math Games Blossom Chart.

▸ Have students participate in several practice rounds.

Materials

✓ MATH GAMES BINDER (SEE CHAPTER 3 FOR DETAILS)

✓ DECK OF CARDS

Lesson Outline

CLASS DISCUSSIONS

▸ What are math games?

▸ Why are math games important?

▸ What can math games help us do?

▸ Let's record our ideas on the Math Games Blossom Chart.

POSSIBLE STUDENT RESPONSES

▸ Games you play to learn math.

▸ Games help you have fun during math.

▸ Math games help us practice math.

▸ Math games help us be better at math.

THE EXPECTATIONS

▸ The only way to get better at math is to spend time practicing what we have learned.

▸ We can have fun while practicing by playing math games!

▸ The goal in fourth grade is to practice our math skills every day.

▸ What things can we do to reach this goal?

THE BLOSSOM CHART

▸ Let's fill out our Blossom Chart.

▸ What should we put in the "Materials" section of our Blossom Chart?

▸ What are the student expectations during Math Games time? What should students be doing? What should our classroom look like? What should our classroom sound like?

▸ How should you treat the materials?

▸ What should you do when it is time to pick up?

▸ What are the teacher expectations during Math Games time? What should the teacher be doing?

POSSIBLE STUDENT RESPONSES

▸ Math Games binder and deck of cards go in the "Materials" section.

▸ Students should be focused and work the whole time.

▸ Students should talk in whisper voices.

▸ Students should sit next to their partners.

▸ Respect all of the materials.

▸ When the round is over, put the materials away the right way.

▸ If you finish a game before the round is over, play a new game.

▸ The teacher is meeting with students.

MODEL

▸ Ask two or more student volunteers to show the class what Math Games practice should look like.

  ▾ Critique the students as they model the Math Games activity.

   I see them sitting next to each other.

   I hear them talking in whisper voices.

   I see them respecting the materials.

   I see them staying focused and working the whole time.

▸ *What did you see (students' names) doing right?*

▸ *Did (students' names) follow all of the expectations on our Blossom Chart?*

POSSIBLE STUDENT RESPONSES

▸ I saw (students' names) working the whole time.

▸ I saw (students' names) talking in whisper voices.

▸ I think (students' names) could have stayed focused a little more.

▸ I like how (students' names) picked up the materials when they were done.

PRACTICE

▸ Invite students to practice a round of Math Games time.

▸ Call students one pair at a time to choose a spot in the classroom to work.

▸ Once all students are working, start timing.

▸ Stop the round if any student is off task or not following the Blossom Chart expectations.

▸ Signal to students that the round is over.

▸ Have students practice picking up the game quickly and quietly and return to the classroom meeting area.

REVIEW

▸ *How did that round go?*

▸ *What items on our Blossom Chart did you see happening?*

▸ *Is there anything we need to work on?*

▸ *Is there anything we need to add to the Blossom Chart?*

▸ Complete another practice round and review if time permits.

POSSIBLE STUDENT RESPONSES

▸ I think we could have worked a little more quietly.

▸ I think we did a good job staying focused.

▸ I think we need to try to come back to the meeting area a little faster.

▸ We should write on the Blossom Chart that you shouldn't sit very close to other groups because they might distract you.

Mathematical Practice

The famous NFL football coach Vince Lombardi once said, "Practice does not make perfect. Perfect practice makes perfect." We believe this is also true in math. The Mathematical Practice component of the Common Core 4 gives students plenty of opportunity to practice their math skills. It also provides time to conduct conferences with individual students to make sure they are practicing how to solve math problems. In order for students to master the Common Core Standards, they need to be provided with adequate time for perfect practice.

> "Practice does not make perfect. Perfect practice makes perfect."
> — Vince Lombardi

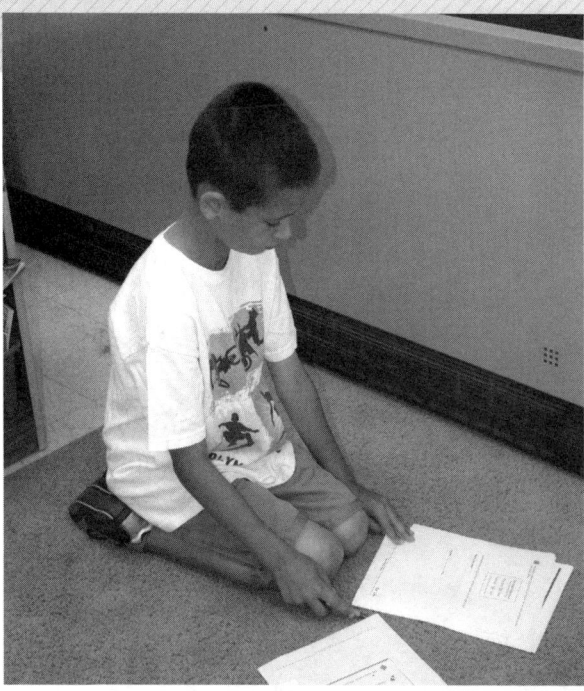

After hearing of the Common Core Standards, many teachers found their old teaching manuals and made copies of worksheets in order to drill their students. However, this goes against the constructivist approach that the Common Core Standards call for. There is a need to offer students an opportunity to practice problems they have been learning about. However, it is equally important to give students the opportunity to use their math knowledge on real-world problems they have never seen before.

The most important part of your Mathematical Practice time will be to provide problem-solving opportunities for your students. This can be accomplished in a variety of ways. It can be a daily math journal in which students record new learning, a problem of the day in which they solve a problem they have never seen using the skills that have been previously taught, or a traditional worksheet. If assigning worksheets, make sure that students are able to work on them independently. The worksheets should focus on skills that have been taught so students are simply reviewing them. It is important to differentiate the Mathematical Practice worksheets. Have worksheets available for your advanced, on-level, and struggling students. This will allow you to challenge the higher-level students without frustrating the lower-level ones.

You might want to include "Counting Clubs" in the Mathematical Practice portion of the Common Core 4. Counting Clubs are number chart worksheets in which students have to count or skip count to a specific number. For students who need a challenge, provide a blank chart with random numbers filled in and have students fill in the missing numbers on

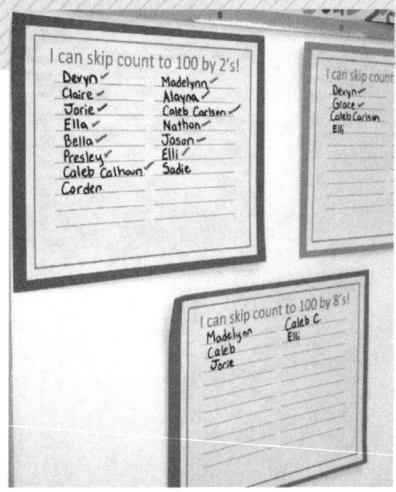

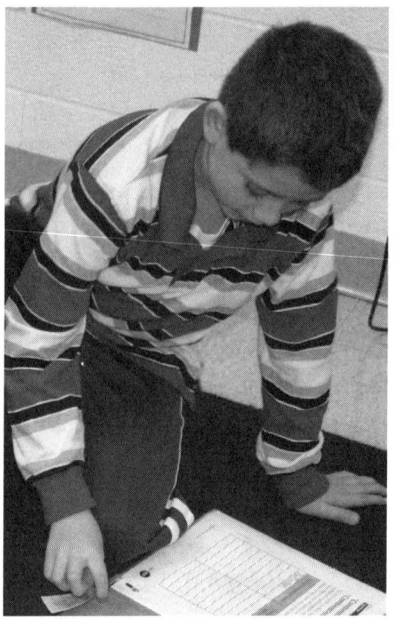

the chart. Or have students skip count by fractions or decimals.

Introduce Mathematical Practice time by bringing the whole class together at the classroom meeting area. Explain to students that they will be learning about the Mathematical Practice activity in the Common Core 4. Ask students to brainstorm reasons why mathematical practice would be an important part of their math learning. Record students' ideas on the Mathematical Practice Blossom Chart.

Then explain to students how Mathematical Practice activities will be set up in the classroom. You can set them up in whatever manner makes the best sense for you. This is really a matter of personal preference. You may choose to create packets for students to keep in their math folders or have baskets or bins of different Mathematical Practice worksheets that students help themselves to during the Mathematical Practice round. You could also set up a math box with a Mathematical Practice folder and math journal for every student in your classroom (see Chapter 3 for details). If you do this, you can provide reference materials, such as a number chart.

Should you decide to compile Mathematical Practice packets for each unit that you will teach, be sure to include a variety of worksheets—both drills and problem solving. Also include worksheets that are specific to the skills taught in the unit and worksheets that reinforce skills that have been taught in previous units. Distribute new packets at the beginning of each unit. No matter how you decide to organize the Mathematical Practice worksheets, allow your students some time to preview the worksheets and your means of organizing them. Once again, allowing time to explore the materials now will minimize distractions later on. After students have been given enough time to look through the materials, it is time to finish the Mathematical Practice Blossom Chart.

As always, students should drive the class discussion with you guiding them to include certain aspects of the routine. As you create the Blossom Chart, be sure to include the expectation that students should work quietly and independently the entire time. It is important for you to address what students should do if they come to a problem on a Mathematical Practice worksheet that they do not know

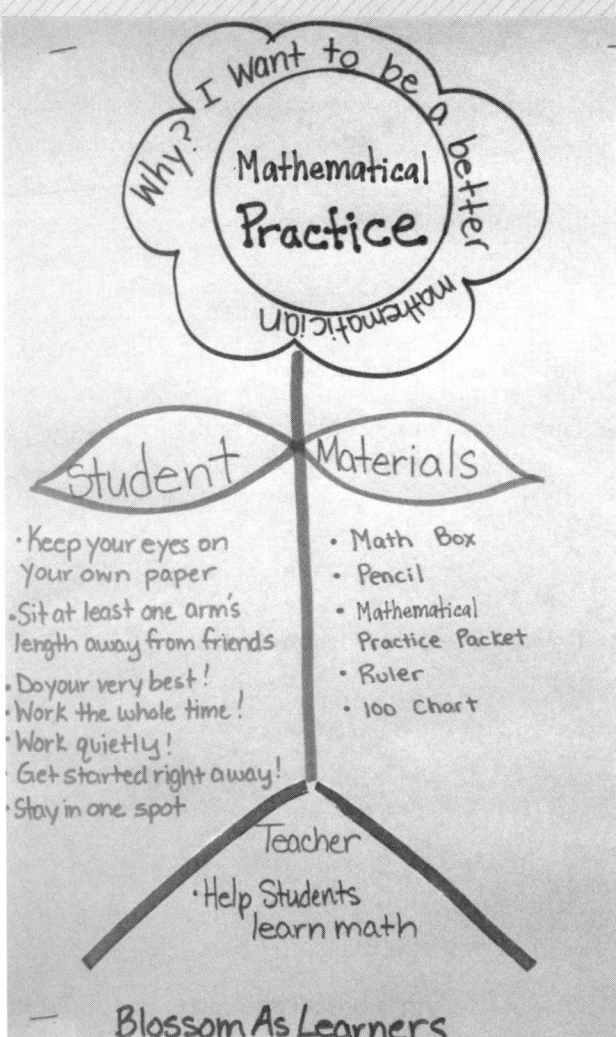

should observe as you critique the student out loud according to the Blossom Chart.

Next, it is time for a practice round. Allow students to sit anywhere in the classroom where they would like to work. Remind them to make a smart choice as to where they sit. Have students begin with a 1- to 2-minute round. While students work, monitor their behavior without engaging them. If any students try to get your attention because they forgot what to do when they come to a problem they cannot attempt, stop the round, bring the class back together, and discuss the strategy for that situation.

Once this round is finished, signal to the class to come back to the classroom meeting area. Discuss what went well and what needs to be worked on. Make any changes or additions to the Blossom Chart. If time permits, attempt another round of Mathematical Practice time.

The second and subsequent days look the same as the first day. Review the Blossom Chart. Have students model correct behavior. Practice for a few rounds. Add 1–2 minutes each day as you see fit. You should be working toward 5 to 15 minutes per round depending on your students and the amount of time you have available for your math block.

how to solve. One suggestion is to have students circle the problem and then skip it. Students should continue working on other problems until you have a conference with them. Once you come to them for a conference, they can quickly look back over their papers to see which problems they needed help with. This is a great way for students to take ownership of their learning during math conferences.

Once you have completed the Blossom Chart, it is time to have several students model the appropriate behaviors for Mathematical Practice time. While each student models, the rest of the class

Weekly Plans

MATHEMATICAL PRACTICE

Day 1

1. Whole-class discussion on importance of mathematical practice
2. Brainstorm why mathematical practice is important
3. Explore Mathematical Practice materials
4. Create Mathematical Practice Blossom Chart
5. Students model Mathematical Practice time
6. Practice round for 1–2 minutes
7. Review and edit Blossom Chart
8. One or two more practice rounds for 1–2 minutes each

Day 2

1. Review Mathematical Practice Blossom Chart
2. Students model mathematical practice
3. Practice round for 2–3 minutes
4. Review and edit Blossom Chart
5. One or two more practice rounds for 2–3 minutes each

Day 3

1. Review Mathematical Practice Blossom Chart
2. Students model mathematical practice
3. Practice round for 3–4 minutes
4. Review and edit Blossom Chart
5. One or two more practice rounds for 3–4 minutes each

Day 4–10

1. Review Mathematical Practice Blossom Chart
2. Practice round: Add 1–2 minutes each day
3. Review and edit Blossom Chart as needed
4. One or two more practice rounds if time permits

Day 11–Beyond

1. Review Mathematical Practice Blossom Chart as needed
2. One full round of Mathematical Practice time for 5 to 15 minutes
3. Review and edit Blossom Chart as needed

Day 1 Lesson Plan

MATHEMATICAL PRACTICE

Objectives

▸ Introduce Mathematical Practice time to students.

▸ Allow students to familiarize themselves with materials.

▸ Create Mathematical Practice Blossom Chart.

▸ Have students participate in several practice rounds.

Materials

✓ MATHEMATICAL PRACTICE PACKET

Lesson Outline

CLASS DISCUSSION

▸ What do you think mathematical practice is?

▸ Why is it important to practice math?

▸ What can mathematical practice help us do?

▸ Let's record our ideas on the Mathematical Practice Blossom Chart.

POSSIBLE STUDENT RESPONSES

▸ Mathematical practice is when we practice types of math that we have been learning.

▸ We can get better at math by practicing.

▸ Mathematical practice helps us remember what we have been learning.

THE EXPECTATIONS

▸ The goal in fourth grade is to practice what we learn in math every day.

▸ What things can we do to reach this goal?

▸ The only way to get better at math is to spend time practicing what we have learned!

▸ We will have time each day for you to practice math.

THE BLOSSOM CHART

▸ Let's fill out our Blossom Chart.

▸ What should we put in the "Materials" section of our Blossom Chart?

▸ What are the student expectations during Mathematical Practice time? What should students be doing? What should our classroom look like? What should our classroom sound like?

▸ What should you do if there is a problem you do not know how to solve?

▸ What are the teacher expectations during Mathematical Practice time? What should the teacher be doing?

POSSIBLE STUDENT RESPONSES

▸ Mathematical Practice packet and pencils go in the "Materials" section.

▸ Students should be focused and work the whole time.

▸ Students should work quietly.

▸ Students should sit far away from friends.

▸ Skip the problem you don't know how to do.

▸ The teacher is meeting with students.

MODEL

- Ask a student volunteer to show the class what Mathematical Practice time should look like.
  - Critique the student as he or she models Mathematical Practice time.
    > *I see (student name) getting started right away.*
    >
    > *I see (student name) working quietly.*
    >
    > *I see (student name) sitting far away from friends.*
    >
    > *I see (student name) staying focused and working the whole time.*
- *What did you see (student name) doing right?*
- *Did (student name) follow all of the expectations on our Blossom Chart?*

POSSIBLE STUDENT RESPONSES

- I saw (student name) working the whole time.
- I saw (student name) talking in a whisper voice.
- I think (student name) could have stayed focused a little more.
- I like how (student name) picked up when he or she was done.

PRACTICE

- Invite students to practice a round of Mathematical Practice time.
- Call students one at a time to choose a spot in the classroom to work.
- Once all students are working, start timing.
- Stop the round if any student is off task or not following the Blossom Chart expectations.
- Signal to students that the round is over.
- Have students practice picking up their materials quickly and quietly and return to the classroom meeting area.

REVIEW

- *How did that round go?*
- *What items on our Blossom Chart did you see happening?*
- *Is there anything we need to work on?*
- *Is there anything we need to add to the Blossom Chart?*
- Complete another practice round and review if time permits.

POSSIBLE STUDENT RESPONSES

- I think we could have worked a little more quietly.
- I think we did a good job staying focused.
- I think we need to try to come back to the meeting area a little faster.

Technology

We have found that instructional technology is an amazing motivator for students. When we use technology in a lesson, our students are not only engaged, but they are begging for the chance to be able to explore their learning through the technology. Because of this, we try to use technology in our classroom every day and as much as possible. We knew we had to make technology one of the Common Core 4 choices.

Ozel et al. (2008) discuss the effects of instructional technology in the math classroom. If used appropriately, instructional technology in the math classroom can increase motivation in students, improve attitudes toward learning, and, most importantly, raise student achievement. If your school has limited access to instructional technology, you can still include technology as part of your Common Core 4. Use whatever technology is available to you during the Technology activity round of Common Core 4 in your classroom. You do not need to have every student in your classroom participating in the Technology activity round at one time. The Technology activity round can be limited to whatever you have available for students to use. If you only have two computers in your classroom, then only two students per round can select Technology time as their choice.

> **We have found that instructional technology is an amazing motivator for students.**

Introduce Technology time in the same sequence as the other Common Core 4 components. Begin by having a whole-class discussion on the purpose of technology during math instruction. Ask students to brainstorm why they would want to use technology during math and how it can help them become better mathematicians. Students always say "We use technology because it is fun!"; try to guide students to

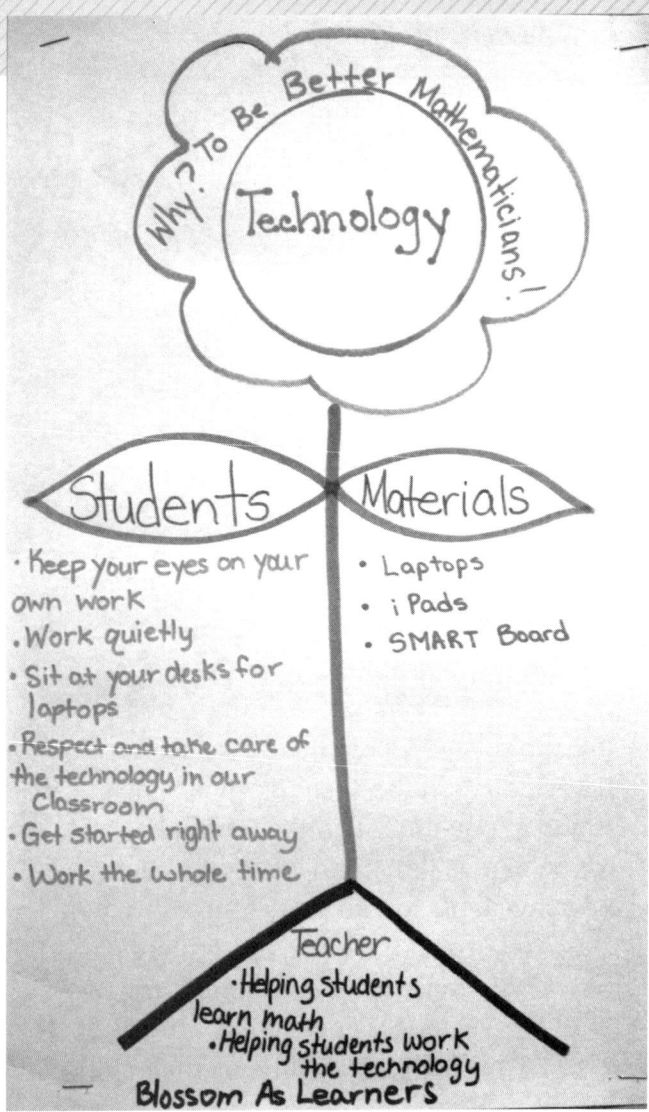

recognize that technology is a big part of the world they are growing up in and will most likely play a major role in their future careers. Record the student ideas and reasoning on the Technology Blossom Chart.

Next, explain to students how they will be using technology to become better mathematicians. Then describe what forms of technology are available to them. Begin to fill out the Blossom Chart for Technology time by listing the technology under the materials section. Continue filling out the Blossom Chart by asking students what the classroom should look like and sound like during Technology time. Allow students to brainstorm and share any ideas they have. Guide the discussion to focus on how to care for the technology that will be used when it is time to clean up. Make sure that the Blossom Chart reminds students to work quietly, get started right away, and work the whole time. Finish the Blossom Chart by asking students to think of what the teacher should be doing during Technology time and recording their responses.

Allow students to brainstorm and share any ideas they have.

Once the Blossom Chart is complete, it is time to introduce students to one of the types of technology. It is a good idea to introduce one type of technology at a time and focus on it for several days before introducing the next type. You want to make sure students have mastered each type of technology first before a new technology is presented. Allow students a few minutes to explore the technology before it is time for them to listen to directions.

After students have had time to set up the technology and explore it, ask for several volunteers to model the right way to use the technology. While the volunteers model, the rest of the class observes and critiques their behaviors according to the Blossom Chart that was just created. Once the modeling is finished, it is time for students' first practice round of using technology. Begin with a 4- to 5-minute round on the first day. While students work with the technology, monitor their behavior from a distance. When the 4–5 minutes are up, signal to students in the same way you do for the rest of the Common Core 4 activities. Because you will be doing another practice round of Technology time, ask students to leave the materials they were using up and running and come to the meeting area. Once students are at the meeting area, have a class discussion on what went well and what they can do better. Then review and make any changes to the Blossom Chart. If time permits, have students practice two to three more 4- to 5-minute rounds of Technology time. Make sure to have the final round end with plenty of time for you to teach students how to properly turn off the device and store it (if needed) until it will be used again.

> When you are confident that students have mastered that particular type of technology, move on to introducing them to a new type.

The next day will be very similar to the first day. Have students work with the same type of technology. Review the Blossom Chart with students, have them practice a round that is 1 to 2 minutes longer than on the previous day, and then discuss what went well and what needs to be worked on. Repeat these steps several more times on the days to follow. When you are confident that students have mastered that particular type of technology, move on to introducing them to a new type.

Weekly Plans

TECHNOLOGY

Day 1

1. Whole-class discussion on importance of technology
2. Brainstorm how technology can help students learn math
3. Inform students of technology choices during Technology time
4. Create Technology Blossom Chart
5. Introduce one type of technology on Day 1
6. Students model the Technology activity
7. Practice round for 4–5 minutes
8. Review and edit Blossom Chart
9. One or two more practice rounds for 4–5 minutes each

Day 2

1. Review Technology Blossom Chart
2. Students model same technology used on Day 1
3. Practice round for 5–6 minutes
4. Review and edit Blossom Chart
5. One or two more practice rounds for 5–6 minutes each

Day 3

1. Review Technology Blossom Chart
2. Students model same technology used on Day 1
3. Practice round for 6–7 minutes
4. Review and edit Blossom Chart
5. One or two more practice rounds for 6–7 minutes each

Day 4–10

1. Review Technology Blossom Chart
2. Practice round using same technology as Day 1—Add 1–2 minutes each day
3. Review and edit Blossom Chart as needed
4. One or two more practice rounds if time permits

Day 11–Beyond

1. Review Technology Blossom Chart as needed
2. One full round of technology for 10 to 15 minutes
3. Review and edit Blossom Chart as needed
4. Introduce and model new forms of technology throughout the school year

Day 1 Lesson Plan

TECHNOLOGY

Objectives

- Introduce Technology time to students.
- Allow students to familiarize themselves with materials.
- Create Technology Blossom Chart.
- Have students participate in several practice rounds.

Materials

- ✓ COMPUTERS
- ✓ TABLETS
- ✓ INTERACTIVE WHITEBOARD

Lesson Outline

CLASS DISCUSSION

- What do you think technology is?
- Why is it important to practice math with technology?
- What can technology help us do?
- Let's record our ideas on the Technology Blossom Chart.

POSSIBLE STUDENT RESPONSES

- Technology is when we use things like the computers.
- We can get better at math by practicing it with technology.
- Technology helps us have fun practicing math.

THE EXPECTATIONS

- The goal in fourth grade is to practice what we learn in math every day.
- What things can we do to reach this goal?
- The only way to get better at math is to spend time practicing what we have learned!

- We will have time each day for you to practice math with technology.

THE BLOSSOM CHART

- Let's fill out our Blossom Chart.
- What should we put in the "Materials" section of our Blossom Chart?
- What are the student expectations during Technology time? What should students be doing? What should our classroom look like? What should our classroom sound like?
- How should you treat the technology?
- What are the teacher expectations during Technology time? What should the teacher be doing?

POSSIBLE STUDENT RESPONSES

- Computers, tablets, and the interactive whiteboard go in the "Materials" section.
- Students should be focused and work the whole time.
- Students should work quietly.
- Students should treat the technology with respect.
- The teacher is meeting with students.

MODEL

▸ Ask a student volunteer to show the class what Technology time should look like.

  ▿ Critique the student as he or she models the Technology activity.

   | *I see (student name) getting started right away.*

   | *I see (student name) working quietly.*

   | *I see (student name) using the technology correctly.*

   | *I see (student name) staying focused and working the whole time.*

▸ *What did you see (student name) doing right?*

▸ *Did (student name) follow all of the expectations on our Blossom Chart?*

POSSIBLE STUDENT RESPONSES

▸ I saw (student name) working the whole time.

▸ I saw (student name) working quietly.

▸ I think (student name) could have stayed focused a little more.

▸ I like how (student name) got started right away.

PRACTICE

▸ Invite students to practice a round of Technology time.

▸ Call students one at a time to choose a spot in the classroom to work (if they are not already at the technology).

▸ Once all students are working, start timing.

▸ Stop the round if any student is off task or not following the Blossom Chart expectations.

▸ Signal to students that the round is over.

▸ Have students return to the classroom meeting area. (For today, have students leave the technology turned on and ready to go for another practice round.)

REVIEW

▸ *How did that round go?*

▸ *What items on our Blossom Chart did you see happening?*

▸ *Is there anything we need to work on?*

▸ *Is there anything we need to add to the Blossom Chart?*

▸ Complete another practice round and review if time permits.

POSSIBLE STUDENT RESPONSES

▸ I think we could have worked a little more quietly.

▸ I think we did a good job staying focused.

▸ I think we need to try to come back to the meeting area a little faster.

▸ We should write on the Blossom Chart that you shouldn't sit next to your friends because they might distract you.

STUDENT CHOICE LOG
COMMON CORE 4

STUDENT NAMES

ROUNDS

MONDAY
1
2
3

TUESDAY
1
2
3

WEDNESDAY
1
2
3

THURSDAY
1
2
3

FRIDAY
1
2
3

Key: **F** MATH FLUENCY **G** MATH GAMES **P** MATHEMATICAL PRACTICE **T** TECHNOLOGY

Teacher Conferences

The mini-lesson followed by independent practice format is effective because it allows time to work with students individually or in small groups. Once routines are in place for each of the Common Core 4 choices and students can demonstrate at least 3 minutes of stamina for independent practice, you can begin to conduct short, informal conferences. Once students are able to work independently for about 5 to 7 minutes, you can begin to hold full conferences with students. This is the time in which students will begin to add and move their names on the GNOMe board according to the skills they have yet to master.

Each conference should last for 4 to 6 minutes for individual meetings and 7 to 10 minutes for small group meetings. Sit next to the students wherever they may be working in the classroom. What you will do during the conference depends on the needs of the students. Sometimes you may simply oversee what the students are working on and help out where appropriate. Other times, you may go to students with a plan of action. If there is a particular skill that you know students are struggling with, spend the conference time reteaching the skill. You do not need to meet with every student or every group every day. You will want to meet with your lower-level students on a consistent, almost daily basis. Higher-achieving students are able to be successful with less frequent math conferences.

When you meet with students, bring your math conference binder. This binder should include a list of all your students' names and space to write the dates you meet with each student. This allows you to keep track of who you are meeting with and when. Divide your math conference binder into sections labeled with students' names. In each section have a copy of the Standards Reference chart (page 51), which has wording similar to the phrases posted on the Math GNOMe bulletin board. It is important to have the standards readily available to reference during your conferences. Another option to using the Standards Reference Chart is to use the bulletin board phrases or *I Can...* Statements reference charts. The various forms provide you the option of using whichever format works best for you and your students. We suggest you also have students keep a copy of the form you choose to use in their math folders. If you determine that a student has mastered a specific standard, invite that student to highlight it on his or her individual reference chart inside his or her math folder. This creates a visual reminder for students to see all of the skills they have learned during the year. Give the student a sticky note, an item of your choosing, or a leaf from the companion décor set with his or her name or assigned class number on it. Then have the student place his or her piece next to a new focus skill on the Math GNOMe bulletin board.

In addition to the standards reference sheet of your choosing, include several copies of the Math Conference Observation Log (page 52) in your math conference binder. Use these reproducibles to keep track of what you do with students during their conference time. Record each student's strengths and weaknesses, as well as what activities you have done with the student during the conference. Write down the next steps for the student to achieve mastery of a particular skill. Discuss with students what they feel are their strengths and weaknesses and what math skills they should focus on during the rounds of the Common Core 4.

Standards Reference Chart
FOURTH GRADE COMMON CORE

Geometry

- Recognize and identify lines 4.G.A.1
- Recognize and identify line segments 4.G.A.1
- Recognize and identify rays 4.G.A.1
- Recognize and identify angles (acute, obtuse, and right) 4.G.A.1
- Recognize and identify perpendicular lines 4.G.A.1
- Recognize and identify parallel lines 4.G.A.1
- Recognize and identify right triangles 4.G.A.2
- Classify shapes by properties of their lines and angles 4.G.A.2
- Recognize, identify, and draw lines of symmetry 4.G.A.3

Number Sense
(NUMBER & OPERATIONS IN BASE TEN AND NUMBER & OPERATIONS—FRACTIONS)

- Recognize a number in one place represents ten times what it is in the place to the right 4.NBT.A.1
- Read and write multi-digit whole numbers 4.NBT.A.2
- Compare two multi-digit numbers using >, <, or = 4.NBT.A.2
- Round multi-digit whole numbers to any place 4.NBT.A.3
- Fluently add multi-digit numbers 4.NBT.B.4
- Fluently subtract multi-digit numbers 4.NBT.B.4
- Multiply up to a 4-digit number by a 1-digit number and two 2-digit numbers 4.NBT.B.5
- Use equations, arrays, and area to explain multiplication and division 4.NBT.B.5
- Find whole number quotients and remainders 4.NBT.B.6
- Recognize, generate, and explain equivalent fractions 4.NF.A.1, 4.NF.C.5
- Compare fractions with <, =, > 4.NF.A.2
- Understand what a fraction is 4.NF.B.3
- Add and subtract fractions 4.NF.B.3a, 4.NF.C.5
- Decompose a fraction into a sum of fractions 4.NF.B.3b
- Add and subtract mixed numbers 4.NF.B.3c
- Solve word problems involving fractions 4.NF.B.3d, 4.NF.4c
- Multiply fractions by whole numbers 4.NF.B.4, 4.NF.B.4a, 4.NF.B.4b
- Understand decimal notation for fractions 4.NF.C.6
- Express a fraction with a denominator of 10 as a fraction with a denominator of 100 4.NF.C.5
- Compare two decimals with >, <, or = 4.NF.C.7

Operations
(OPERATIONS & ALGEBRAIC THINKING)

- Interpret multiplication equations as comparisons 4.OA.A.1
- Multiply or divide to solve word problems 4.OA.A.2
- Use the four operations (+, -, ×, ÷) to solve multistep word problems 4.OA.A.3
- Solve problems in which remainders must be interpreted 4.OA.A.3
- Use a letter to stand for an unknown quantity 4.OA.A.3
- Find all factor pairs in the range of 1–100 4.OA.B.4
- Recognize a whole number is a multiple of its factors 4.OA.B.4
- Determine whether a number is prime or composite 4.OA.B.4
- Determine if a whole number is a multiple of a given number 4.OA.B.4
- Generate a number or shape pattern that follows a given rule 4.OA.C.5
- Identify features of patterns 4.OA.C.5

Measurement
(MEASUREMENT & DATA)

- Know relative sizes of measurement units 4.MD.A.1
- Record measurement equivalents in two-column tables 4.MD.A.1
- Use the four operations (+, -, ×, ÷) to solve word problems involving measurement 4.MD.A.2
- Apply the perimeter formula 4.MD.A.3
- Apply the area formula 4.MD.A.3
- Create a line plot to display a data set involving fractions 4.MD.B.4
- Use and interpret information on a line plot 4.MD.B.4
- Recognize angles 4.MD.C.5
- Measure angles in whole number degrees 4.MD.C.5a, 4.MD.C.5b, 4.MD.C.6
- Recognize angle measure as additive 4.MD.C.7
- Solve addition and subtraction problems to find unknown angles 4.MD.C.7

Observation Log
MATH CONFERENCE

Common Core 4

Student Name:

DATE: FOCUS STRATEGY/SKILL:

OBSERVATION: INSTRUCTION:

GOAL FOR NEXT MEETING:

DATE: FOCUS STRATEGY/SKILL:

OBSERVATION: INSTRUCTION:

GOAL FOR NEXT MEETING:

DATE: FOCUS STRATEGY/SKILL:

OBSERVATION: INSTRUCTION:

GOAL FOR NEXT MEETING:

Conferences Overview

Individual Student

1 MINUTE

Observe student working on individual math activities such as a Common Core 4 choice. Record your observations.

1–2 MINUTES

Comment on strengths and weaknesses observed. Review the student's focus skill or strategy and the goal that was set at the last meeting.

Discuss what the student has done to reach this goal and work toward mastering the skill or strategy.

1–2 MINUTES

Provide one-on-one instruction in the area of need for the student. Record what instruction was given.

1 MINUTE

Set a new goal with the student for the next meeting. Change the student's focus skill or strategy if need be. (Once a strategy is mastered, a new strategy will become the focus.)

Small Group

1 MINUTE

Gather students that are working on a similar math skill or strategy in a common meeting area in the classroom.

1–2 MINUTES

Review with students what the skill or strategy is and why it is an important part of the math curriculum.

4–5 MINUTES

Provide small-group instruction in the area of need for the group of students.

1–2 MINUTES

Review the instruction that was provided and set a new goal for students to work toward until the next group meeting.

Conducting a Conference
INDIVIDUAL STUDENT

1 Minute

- Review the standard the student is working on.
- Observe the student working on his or her Common Core 4 choice.

> **POSSIBLE TEACHER DIALOGUE:**
>
> *What standard have you been working on?*
> *How do you feel you have been doing with that standard?*
> *Is there anything I can help you with?*

2–4 Minutes

- Based on your observations, provide direct instruction to the student regarding the standard he or she is working on.

> **POSSIBLE TEACHER DIALOGUE:**
>
> *I see you skipped this problem. Can we work together to try to solve it?*
> *Let me show you a strategy that might help you master the standard.*
> *Think back to our lesson on this standard. Try solving the problem the same way we did at the interactive whiteboard.*

1 Minute

- Set a goal and steps to take before the next conference.

> **POSSIBLE TEACHER DIALOGUE:**
>
> *Do you feel you have mastered this skill or do you want to continue working on it?*
> *I think you need a few more days to work on this standard. I will meet with you again on Thursday.*
> *You seem pretty confident with this standard. Let's find a new standard for you to begin working on.*

Conducting a Conference
SMALL GROUP

1–2 Minutes

▸ Review the standard the students are working on.
▸ Discuss the importance of this skill and why it is an important part of the Common Core Standards.

> **POSSIBLE TEACHER DIALOGUE:**
>
> *What standard have you been working on?*
>
> *Why is this standard important for you to know?*

4–5 Minutes

▸ Provide direct instruction to the students regarding the standard they are working on.

> **POSSIBLE TEACHER DIALOGUE:**
>
> *Let me show you a strategy that might help you master the standard.*
>
> *Think back to our lesson on this standard. Try solving these problems the same way we did at the interactive whiteboard.*

1–2 Minutes

▸ Set a goal and steps to take before the next conference.

> **POSSIBLE TEACHER DIALOGUE:**
>
> *Do you feel you have mastered this skill or do you want to continue working on it?*
>
> *I think you need a few more days to work on this standard. I will meet with all of you again on Thursday.*
>
> *You seem pretty confident with this standard. Should we find a new standard for the group to begin working on?*

Sample Conference

The following is a sample conference with a fourth grade student who is currently working on fluently adding and subtracting multi-digit whole numbers using the standard algorithm.

TEACHER:
Sarah, the last time we had a conference together, we agreed that subtraction using the standard algorithm would be a great standard for you to work on. Last time, you were having trouble regrouping across two place values, like in 500 – 157. How has it been going?

SARAH:
Not so good. I still get tripped up with the regrouping. Can you help me?

TEACHER:
Let's try answering the problem 500 – 157 while using the place value blocks. First, write down the problem 500 – 157, and then show me the number 500 using the place value blocks.

SARAH:
Sure! (Writes down the problem and shows 5 of the 100 place value blocks.)

TEACHER:
Great job! How many ones do we need to subtract from 500?

SARAH:
Well, I know I need to take away 7 but I do not have 7 ones, and I know I need to regroup, but this is where I get stuck.

TEACHER:
OK. Let's work through this together. If I had the number 500 and I took 1 away, what would I have?

SARAH:
499.

TEACHER:
Great. Show me 499 using blocks. (Sarah shows 4 hundred, 9 tens, and 9 ones.) Great, Sarah. Now add that 1 back into the ones place and tell me what you see.

SARAH:
I see 4 hundreds, 9 tens, and 10 ones.

TEACHER:
Is that the same as the number 500 only regrouped?

SARAH:
Yes.

TEACHER:
OK. Now cross off 500 and rewrite it as 4 hundreds, 9 tens, and 10 ones. Can you take 7 ones away from the 10 ones?

SARAH:
YES! I can also take away the 5 tens from 9 tens and the 1 hundred from the 4 hundreds for an answer of 343!

TEACHER:
Great! Try 600 – 245. If you get stuck, please use the manipulatives, but remember your goal is to learn the standard algorithm, so focus on place value.

SARAH:
OK, I will!

CHAPTER 3
Preparing Your Classroom

The biggest challenge to getting started is organizing all of your math materials.

The way you choose to organize your materials is really up to you; use your preferred system or procedure. There is no one way that will work perfectly for every teacher. In this chapter we will share with you how we organized our classrooms.

Bulletin Boards

Dedicate a large bulletin board in your classroom to the Math GNOMe. Divide this bulletin board into four sections. Label the sections *Geometry, Number Sense, Operations*, and *Measurement*. In each of the four sections, leave room to add the standards. Do not post all of the standards on the first day of school. Rather, gradually add to the bulletin board throughout the school year as you introduce each one. Make sure that this board is easily accessible to you and your students on a daily basis. Remember to constantly update it as you introduce new skills to your students.

Post the Blossom Charts that you create with your students on a dedicated bulletin

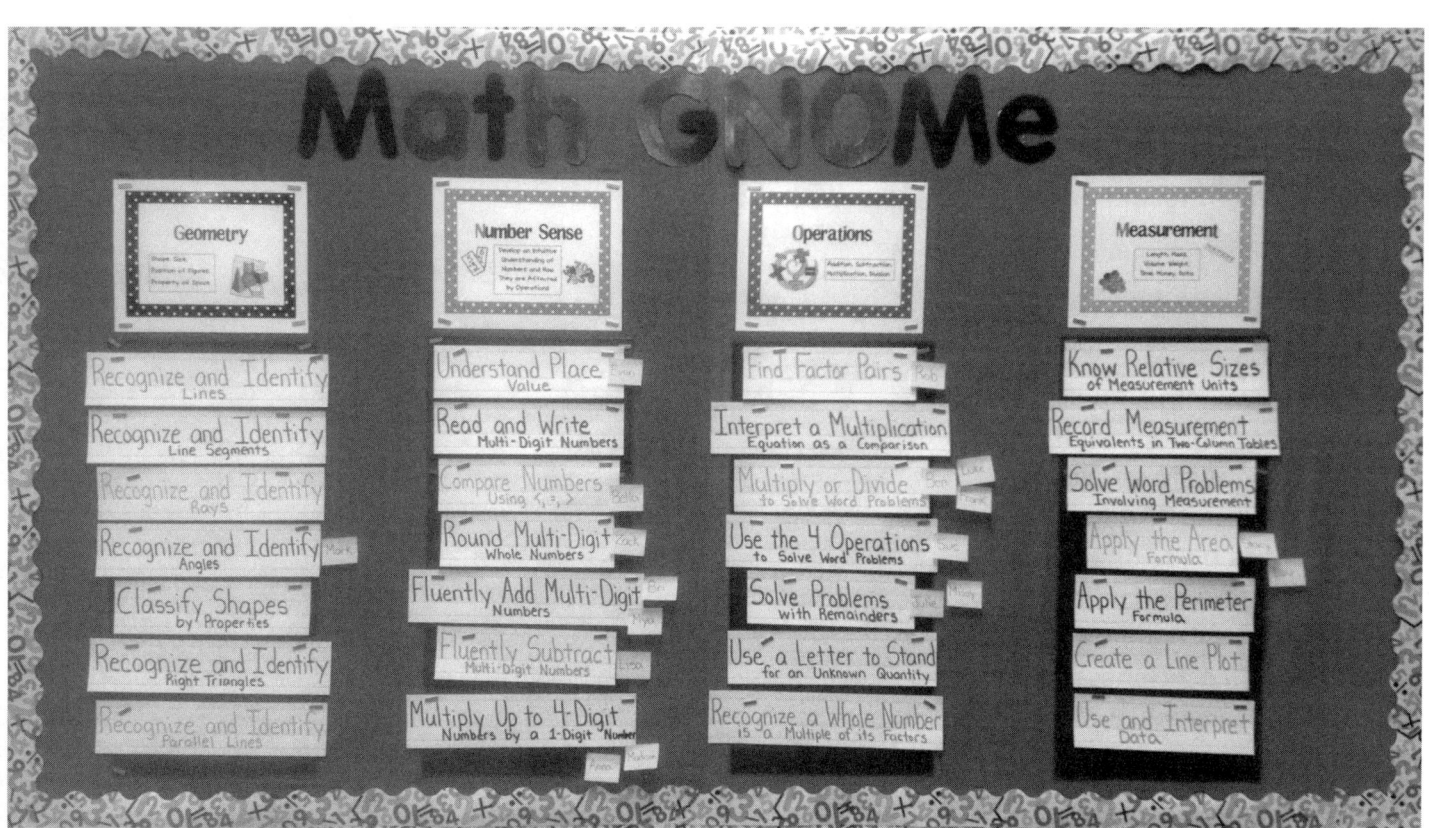

board on or near your Math GNOMe board. In advance, write students' names on cards that can be attached to the Blossom Chart to denote each student's choice of Common Core 4 activity. (See Chapter 2 for Blossom Chart details.) If you have space available, you may also wish to post math references and other helpful charts.

Math Games Binder

Make copies for every game board, game directions, and recording sheets for each student's Math Games activity binder. Another option is to place one copy of each game board, game directions, and recording sheet in a plastic sheet protector for each student. Students can use wipe-off markers to write on the game boards and the recording sheets. Then insert these materials in each student's three-ring binder or two-pocket folder with prongs. This allows you to reuse the binders, game boards, and recording sheets. The binders also make it easy for students to find the materials they need for the game.

Math Boxes

Math boxes are an efficient way to store and organize materials in one place. You can use plastic baskets, small cardboard boxes, and magazine holders. Another organization suggestion is to assign a number to each student according to his or her name. Then write students' numbers on the front of their math boxes. This allows students to quickly locate their math boxes during Common Core 4 rounds. Using numbers as opposed to student names also allows you to reuse the math boxes year after year. The math boxes hold all of the math materials—the

Math Games activity binder, pencils, wipe-off markers, and the Mathematical Practice activity packet.

The Math Games activity binder, Mathematical Practice activity packet, pencils, and wipe-off markers are must-haves in the math box. From there, use your own discretion as to what additional materials you would like students to have immediate access to. You may decide to have your students keep their math facts cards in their boxes. If many of your math games include a deck of numeral cards or a pair of dice, have students store those materials in their math boxes as well.

Below is a suggested list of materials you can place in each student's math box:

Manipulatives

Reserve the math box for the items that students will use on a consistent, almost daily basis. Other materials and manipulatives can be stored in easy-to-access places around your classroom. Make sure that items are clearly labeled and readily available for students. When students choose to play a game or participate in an activity that requires a certain manipulative, they will be able to quickly locate it and get started right away.

Materials

- ✓ MATH GAMES BINDER
- ✓ FOLDER
- ✓ MATHEMATICAL PRACTICE PACKET(S)
- ✓ PENCIL
- ✓ WIPE-OFF MARKER
- ✓ WIPE-OFF MARKER ERASER
- ✓ RULER
- ✓ MATH FACTS CARDS
- ✓ NUMERAL CARDS FOR MATH GAMES
- ✓ DICE

CHAPTER 4
Putting It All Together

> We spend the first five to six weeks of the school year introducing the Math GNOMe and Common Core 4 to our students.

We spend a lot of time establishing the routines during our math block. You will want the expectations and routines to be ingrained in your students so that you do not need to worry about the rest of the class during your math conferences. You will want to be able to completely focus on the student or group of students you are working with. The only way you can do this is if you are confident that the other students are participating in the meaningful learning activities in the correct way the entire time.

When you put it all together and have your mini-lesson and independent activity routine up and running, your daily math block should look like this: Begin by choosing either a new standard to add to the Math GNOMe board or a previously introduced standard to review. Conduct the day's first mini-lesson, which should last 7–10 minutes. Throughout the mini-lesson, reference the Math GNOMe board and make connections to the standards. Then invite students to choose one of the Common Core 4 activities. While students work independently, circulate around the room conducting conferences with individual students and groups of students. This independent time lasts approximately 5–15 minutes, depending on time available. Signal students to return to the classroom meeting area. Start the next mini-lesson, repeating all of the steps above. Require students to select a different activity for each round. Ideally, there should be three mini-lessons and three rounds of the Common Core 4 during each math block.

Keep track of your students' Common Core 4 choices by maintaining a checklist of students' names during each round of Common Core 4 (page 49). After each mini-lesson, call students one by one to make their Common Core 4 choices. As each student selects an activity, write a shorthand notation associated with the activity next to his or her name. For example, *F* stands for Math Fluency, *G* stands for Math Games. Once students have made their selections, have them place their name next to the Common Core 4 activity of their choice (see display shown at right), move to the area of the classroom where they would like to work, and get started right away.

Taking a few minutes to record your students' choices allows you to see what activity each student is working on. It also ensures that students are not doing the same activity every round. You can help guide your students' choices by making recommendations during this check-in. Recording students' choices also allows you to monitor how many students are working on a particular activity at the same time.

The Math GNOMe and Common Core 4 can be used with any math program. There are many ways in which this can be done. One way is to divide your math block in half. During the first half, teach a lesson from your math program. During the second half, teach a Math GNOMe mini-lesson and do one round of the Common Core 4. Or you can keep the format we've described for mini-lesson and independent practice. Try to break your math program's daily lessons into mini-lessons. During the first two mini-lessons, you can teach from the math series. For your third mini-lesson, teach

> The Math GNOMe and Common Core 4 can be used with any math program.

from the Math GNOMe. In between, have your students participate in the Common Core 4 independent practice.

You do not need to use both the Math GNOMe and Common Core 4 in your classroom at the same time. They are independent of each other. You can teach a Math GNOMe mini-lesson followed by the independent practice choices you have developed yourself or gathered from other resources.

On the following pages, you will find examples of different ways to set up your math block using the Math GNOMe and Common Core 4 together, as well as ways to use the Math GNOMe by itself or with other math programs.

Sample Classroom Schedule
USING THE MATH GNOME & THE COMMON CORE 4 EXCLUSIVELY

7–10 Minutes
Math GNOMe Mini-Lesson 1
Teach or review a skill from the Math GNOMe board

5–15 Minutes
Common Core 4 Round 1
Students choose one of the Common Core 4 activities
Teacher meets with students individually or in small groups

7–10 Minutes
Math GNOMe Mini-Lesson 2
Teach or review a skill from the Math GNOMe board

5–15 Minutes
Common Core 4 Round 2
Students choose one of the Common Core 4 activities
Teacher meets with students individually or in small groups

7–10 Minutes
Math GNOMe Mini-Lesson 3
Teach or review a skill from the Math GNOMe board

5–15 Minutes
Common Core 4 Round 3
Students choose one of the Common Core 4 activities
Teacher meets with students individually or in small groups

Sample Classroom Schedule
USING THE MATH GNOME, THE COMMON CORE 4 & MATH SERIES*

7–10 Minutes

Math GNOMe Mini-Lesson 1 or Math Series Mini-Lesson 1

*Teach or review a skill from the Math GNOMe board
or teach a lesson from your math series*

5–15 Minutes

Common Core 4 Round 1

*Students choose one of the Common Core 4 activities
Teacher meets with students individually or in small groups*

7–10 Minutes

Math GNOMe Mini-Lesson 2 or Math Series Mini-Lesson 2

*Teach or review a skill from the Math GNOMe board
or teach a lesson from your math series*

5–15 Minutes

Common Core 4 Round 2

*Students choose one of the Common Core 4 activities
Teacher meets with students individually or in small groups*

7–10 Minutes

Math GNOMe Mini-Lesson 3 or Math Series Mini-Lesson 3

*Teach or review a skill from the Math GNOMe board
or teach a lesson from your math series*

5–15 Minutes

Common Core 4 Round 3

*Students choose one of the Common Core 4 activities
Teacher meets with students individually or in small groups*

* Math series from a major publisher.

Sample Classroom Schedule
USING THE MATH GNOME & SUPPLEMENTAL MATERIALS

7–10 Minutes

Math GNOMe Mini-Lesson 1

Teach or review a skill from the Math GNOMe board

5–15 Minutes

Math Activities

Students participate in math activities from supplemental materials

7–10 Minutes

Math GNOMe Mini-Lesson 2

Teach or review a skill from the Math GNOMe board

5–15 Minutes

Math Activities

Students participate in math activities from supplemental materials

7–10 Minutes

Math GNOMe Mini-Lesson 3

Teach or review a skill from the Math GNOMe board

5–15 Minutes

Math Activities

Students participate in math activities from supplemental materials

Sample Classroom Schedule

USING THE MATH GNOME, A MATH SERIES & SUPPLEMENTAL MATERIALS

7–10 Minutes

Math GNOMe Mini-Lesson 1 or Math Series Mini-Lesson 1

Teach or review a skill from the Math GNOMe board or teach a lesson from your math series

5–15 Minutes

Math Activities

Students participate in math activities from a math series or supplemental materials

7–10 Minutes

Math GNOMe Mini-Lesson 2 or Math Series Mini-Lesson 2

Teach or review a skill from the Math GNOMe board or teach a lesson from your math series

5–15 Minutes

Math Activities

Students participate in math activities from a math series or supplemental materials

7–10 Minutes

Math GNOMe Mini-Lesson 3 or Math Series Mini-Lesson 3

Teach or review a skill from the Math GNOMe board or teach a lesson from your math series

5–15 Minutes

Math Activities

Students participate in math activities from a math series or supplemental materials

Scope & Sequence
COMMON CORE 4

Introduce
Introduce then practice the activity each day.

Continue
Continue to practice, adding 1–2 minutes each day.

Full round
Full round (5–15 min.) of the activity each day.

WEEKS IN THE SCHOOL YEAR	COMMON CORE 4 PLANS
Week 1	**Introduce** Math Fluency
Week 2	**Continue** Math Fluency **Introduce** Math Games
Week 3	**Full round** of Math Fluency **Continue** Math Games **Introduce** Mathematical Practice
Week 4	**Full round** of Math Fluency **Full round** of Math Games **Continue** Mathematical Practice **Introduce** Technology
Week 5	**Full round** of Math Fluency **Full round** of Math Games **Full round** of Mathematical Practice **Continue** Technology
Week 6 and Beyond	**Full round** of Math Fluency **Full round** of Math Games **Full round** of Mathematical Practice **Full round** of Technology

CHAPTER 5
Assessment

Assessment plays a major role in the Math GNOMe and Common Core 4.

Assessment can be seen as the driving force of math instruction during math conferences. There are two types of assessment that teachers use in order to assess students' learning. The first type is summative assessment, which includes end of the unit assessments and standardized testing. The second type, formative assessment, is defined as assessment used to promote student learning success (Stiggins 2005). It is assessing children more frequently for the purpose of improving instructional practices. A more student-empowering approach to formative assessment is "assessment for learning." The Math GNOMe and Common Core 4 offer opportunities for both formative assessment and assessment for learning.

> With proper instruction and assistance from teachers, students will be able to make informed decisions about their own learning.

The difference between formative assessment and assessment for learning is that most formative assessment is intended to inform teachers about what students are learning and to help plan future lessons. Assessment for learning informs students about their own learning. It assumes that students are data-based instructional decision makers, too. With proper instruction and assistance from teachers, students will be able to make informed decisions about their own learning. When students are involved in their own learning and are able to take ownership of their progress, student achievement is greatly increased.

Summative Assessment

Summative assessment helps determine where students are in regards to the standards you are teaching. Paul Bambrick-Santoyo (2010) writes about the importance of data-driven instruction. His book *Driven by Data* details the importance of beginning the school year with a pre-assessment to determine the skills that your students are coming in with.

After students have completed a unit of study, give an assessment to see who has met or exceeded the standards and who still needs to master the standards. Bambrick-Santoyo describes the importance of following up a unit or interim assessment a week later by reteaching standards that students had not mastered.

Formative Assessment

Data-driven instruction is one of the big initiatives happening across the country. Small-group and individual instruction allow for this type of assessment. Formative assessment occurs during the individual and small-group meetings of the Common Core 4. During these meetings, time is spent reteaching the skills as well as assessing students' growth within the particular skill. The information obtained is then used to guide instruction.

Assessment for Learning

Rick Stiggins describes assessment for learning as when teachers use "many different assessment methods to provide students, teachers, and parents with a continuing stream of evidence of student progress in mastering the knowledge and skills that underpin or lead up to state standards" (Stiggins 2005). By using the Math GNOMe and the Common Core 4, students are able to visually see the learning standards and what is expected of them, as well as practice them with meaningful activities on a daily basis. Assessment for learning takes place while students are engaged in the meaningful Common Core 4 activities.

During subsequent conferences, the teacher will monitor and assess students' skill performance.

Math conferences provide time for assessment for learning. During this time, the teacher observes students working and records anecdotal notes in the math conference binder that are used to facilitate future conferences. They can also be used to suggest math games that address certain skills. During subsequent conferences, the teacher will monitor and assess students' skill performance. Together they will continue to set goals and discuss steps to attain them, which will be documented on the conference sheet. The student is encouraged to communicate evidence of learning to the teacher during each conference. This evidence is also recorded on the conference sheet.

The most important part of this process of assessment for learning is the ability of students to track their own learning and celebrate mastery of skills with their teacher and their peers. Once a skill has been mastered, it is important to celebrate this achievement with the class and send a note home to the parents (Chapter 6).

CHAPTER 6
Home Connection

Communication between the school and home is an integral part of a student's education.

Research has shown that the more connected a student's support system is to his or her education, the more likely the child is to succeed academically. Hiatt-Michael (2001) discusses the impact that a great home-school relationship can have on a student's progress and attitude in school and toward education in general. It is important to keep the communication between teachers and parents frequent and positive.

"The Common Core State Standards provide a consistent, clear understanding of what students are expected to learn, so teachers and parents know what they need to do to help them. The standards are designed to be robust and relevant to the real world, reflecting the knowledge and skills that our young people need for success in college and careers. With American students fully prepared for the future, our communities will be best positioned to compete successfully in the global economy."*

It is our job as educators to make parents aware of the Common Core Standards, the expectations placed on their children, and the ways they can support learning at home. Teachers should also encourage parents to use the same academic language at home that is used in school and emphasize how their support improves their children's success as lifelong learners.

*Common Core State Standards Mission Statement

When introducing the Math GNOMe and Common Core 4 to your students, also send a letter to parents explaining the key pieces of your math instruction. Throughout the year, update parents on how the Math GNOMe and Common Core 4 are progressing. Include the specific skills or strategies that are being taught from the Math GNOMe.

On the following pages you will find sample home-link letters. Customize the letters to meet your student and classroom needs. As the year progresses, provide parents with new tips and suggestions for ways to practice at home. The more information and ideas you share with your students' families, the more parents will feel they are an integral part of their children's math success.

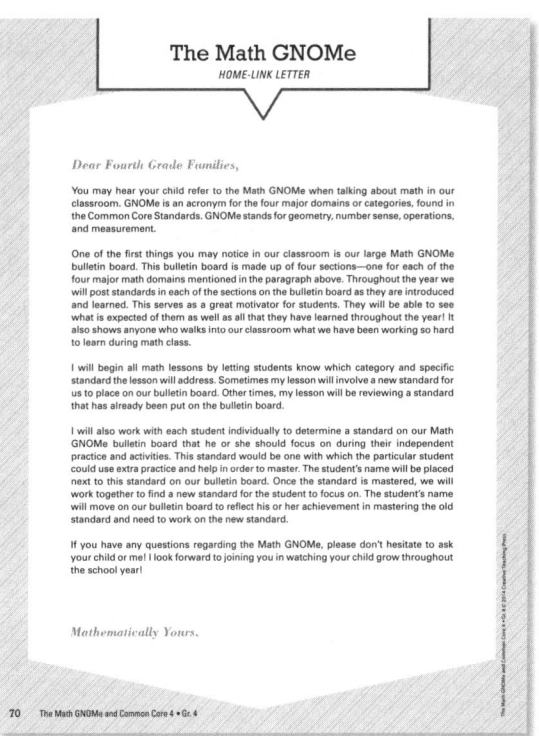

The Math GNOMe
HOME-LINK LETTER

Dear Fourth Grade Families,

You may hear your child refer to the Math GNOMe when talking about math in our classroom. GNOMe is an acronym for the four major domains or categories, found in the Common Core Standards. GNOMe stands for geometry, number sense, operations, and measurement.

One of the first things you may notice in our classroom is our large Math GNOMe bulletin board. This bulletin board is made up of four sections—one for each of the four major math domains mentioned in the paragraph above. Throughout the year we will post standards in each of the sections on the bulletin board as they are introduced and learned. This serves as a great motivator for students. They will be able to see what is expected of them as well as all that they have learned throughout the year! It also shows anyone who walks into our classroom what we have been working so hard to learn during math class.

I will begin all math lessons by letting students know which category and specific standard the lesson will address. Sometimes my lesson will involve a new standard for us to place on our bulletin board. Other times, my lesson will be reviewing a standard that has already been put on the bulletin board.

I will also work with each student individually to determine a standard on our Math GNOMe bulletin board that he or she should focus on during their independent practice and activities. This standard would be one with which the particular student could use extra practice and help in order to master. The student's name will be placed next to this standard on our bulletin board. Once the standard is mastered, we will work together to find a new standard for the student to focus on. The student's name will move on our bulletin board to reflect his or her achievement in mastering the old standard and need to work on the new standard.

If you have any questions regarding the Math GNOMe, please don't hesitate to ask your child or me! I look forward to joining you in watching your child grow throughout the school year!

Mathematically Yours,

The Common Core 4
HOME-LINK LETTER

Dear Fourth Grade Families,

You may hear your child refer to Common Core 4 when talking about math in our classroom. Common Core 4 refers to the four independent practice activities students will engage in on a daily basis. The four choices are Math Fluency, Math Games, Mathematical Practice, and Technology. Each day your child will have the opportunity to work on these four activities to support and enhance his or her math learning.

When your child chooses Math Fluency time, he or she will have the chance to memorize and work toward automaticity with his or her addition, subtraction, multiplication, and division facts. The Common Core Standards place a great emphasis on numeracy fluency. In fourth grade, students are expected to fluently add and subtract multi-digit whole numbers using the standard algorithm. By practicing their math fluency on a daily basis, students will have a great opportunity to master this standard.

Math games are a fun and engaging means of getting students to practice the math skills they have been learning and partake in "math talk" with their peers. During Math Games time, students play games with a partner that support what they have been learning in the classroom.

Mathematical Practice time allows students to use their math knowledge in order to solve new problems. The activities included in Mathematical Practice time are worksheets, math journals, and problem of the day.

During Technology time, students are able to practice their math skills through a variety of instructional technology. Students can use tablets, computers, and the interactive whiteboard. We have access to many great apps and websites that follow the Common Core Standards.

If you have any questions regarding the Common Core 4, please don't hesitate to ask your child or me! I look forward to joining you in watching your child grow throughout the school year!

Mathematically Yours,

Math Fluency
HOME-LINK LETTER

Dear Fourth Grade Families,

We have begun to learn about and practice the Math Fluency portion of Common Core 4! Math Fluency time is one of the four independent practice activities that students will participate in on a daily basis during our math block. The Common Core Standards place a great emphasis on numeracy fluency. In fourth grade, students are expected to fluently add and subtract multi-digit whole numbers using the standard algorithm. By practicing their math fluency on a daily basis, students have a great opportunity to master this standard.

When your child chooses the Math Fluency activity, he or she will have the chance to work with materials within our classroom to memorize and work toward automaticity with his or her math facts. Students have begun to explore these materials and practice what Math Fluency time should look like in our classroom. The students helped to create the expectations, which include working quietly, staying focused, and practicing math facts the whole time.

Ways to Practice at Home

- ✓ Use flash cards, interactive websites, and tablet or smartphone apps to memorize multiplication and division facts.
- ✓ Roll dice and multiply the two numbers.
- ✓ Choose two cards from a deck and multiply the two numbers.
- ✓ Use manipulatives (buttons, macaroni, beads, etc.) to represent equations or create arrays.

If you have any questions regarding Math Fluency time or the Common Core 4, please don't hesitate to ask your child or me! I look forward to joining you in watching your child grow throughout the school year!

Mathematically Yours,

Math Games
HOME-LINK LETTER

Dear Fourth Grade Families,

We have begun to learn about and practice the Math Games time portion of Common Core 4! Math Games time is one of the four independent practice activities that students will participate in on a daily basis during our math block. During Math Games time, students play games that support what we have been learning in the classroom.

Math games are a fun and engaging means of getting students to not only practice the skills they have been learning but to also partake in "math talk" with their peers. Students have begun to explore the game materials and practice what Math Games time should look like in our classroom. Students helped in creating the expectations, which include working quietly, staying focused, being a good sport, and using math vocabulary.

Ways to Practice at Home

- ✓ **Card Capture:** Divide a pile of multiplication and/or division facts cards equally among two players. One player shows the facts side of the card to the opponent. The opponent says the answer to the fact. If the opponent guesses correctly, he or she gets to "capture" the card. This continues until the opponent says an incorrect answer. Then the roles switch.

- ✓ **Concentration:** Write multiplication and/or division facts on one set of cards. Write the answers to the equations on a different set of cards. Set the cards facedown in a grid. Take turns flipping over two cards at a time trying to match equations to their answer.

- ✓ **Fraction Fun:** In this card game, the Ace is worth 11, Jack is worth 12, Queen is worth 13, and the King is worth 14. Divide a deck of cards equally between two players. At the same time, players turn over the top two cards in their piles to create a fraction (the smaller number card is the numerator and the larger number card is the denominator). The student who creates the larger fraction collects all four cards. If the cards create equivalent fractions, the cards are placed in a center pile and regular game play continues. The next hand is played and the player with the larger fraction collects all four cards in addition to the previously played equivalent fraction cards. The game continues until all cards in the piles have been played. The player who collected the most cards wins.

If you have any questions regarding Math Games time or the Common Core 4, please don't hesitate to ask your child or me! I look forward to joining you in watching your child grow throughout the school year!

Mathematically Yours,

Mathematical Practice
HOME-LINK LETTER

Dear Fourth Grade Families,

We have begun to learn about and practice the Mathematical Practice activity portion of Common Core 4! Mathematical Practice time is one of the four independent practice activities that students will participate in on a daily basis during our math block. Mathematical Practice time allows students to use their math knowledge in order to solve new problems. The activities included in Mathematical Practice time are worksheets, math journals, and problem of the day.

Ways to Practice at Home

- ✓ **Problem of the Day**—Create a math problem for your child to solve.
- ✓ **Number of the Day**—Choose a number and have your child come up with multiplication and division equations to make that number.
- ✓ **Math Journal**—Have your child write about when he or she has seen or used math in the real world.
- ✓ **Grocery Store Math**—Have your child mentally add prices, calculate change, and practice counting money when shopping.

If you have any questions regarding Mathematical Practice time or the Common Core 4, please don't hesitate to ask your child or me! I look forward to joining you in watching your child grow throughout the school year!

Mathematically Yours,

Technology
HOME-LINK LETTER

Dear Fourth Grade Families,

We have begun to learn about and practice the Technology activity portion of Common Core 4! Technology time is one of the four independent practice activities that students will participate in on a daily basis during our math block. Students are able to practice their math skills through a variety of instructional technology.

When your child chooses Technology time, he or she will have the opportunity to use tablets, computers, and the interactive whiteboard. We have access to useful apps, websites, and computer programs that follow the Common Core Standards. Students have begun to explore these materials and practice what Technology time should look like in our classroom. The students helped to create the expectations, which include working quietly, staying focused, and treating the technology with respect.

Ways to Practice at Home

- ✓ Interactive websites
- ✓ Tablet apps
- ✓ Educational computer games
- ✓ Smartphone apps

If you have any questions regarding Technology time or the Common Core 4, please don't hesitate to ask your child or me! I look forward to joining you in watching your child grow throughout the school year!

Mathematically Yours,

COMMON CORE STATE STANDARDS

for Mathematical Practice

1 CCSS.Math.Practice.MP1:
Make sense of problems and persevere in solving them.

2 CCSS.Math.Practice.MP2:
Reason abstractly and quantitatively.

3 CCSS.Math.Practice.MP3:
Construct viable arguments and critique the reasoning of others.

4 CCSS.Math.Practice.MP4:
Model with mathematics.

5 CCSS.Math.Practice.MP5:
Use appropriate tools strategically.

6 CCSS.Math.Practice.MP6:
Attend to precision.

7 CCSS.Math.Practice.MP7:
Look for and make use of structure.

8 CCSS.Math.Practice.MP8:
Look for and express regularity in repeated reasoning.

COMMON CORE STATE STANDARDS

for Fourth Grade

Operations & Algebraic Thinking

USE THE FOUR OPERATIONS WITH WHOLE NUMBERS TO SOLVE PROBLEMS.

CCSS.Math.Content.4.OA.A.1
Interpret a multiplication equation as a comparison, e.g., interpret 35 = 5 × 7 as a statement that 35 is 5 times as many as 7 and 7 times as many as 5. Represent verbal statements of multiplicative comparisons as multiplication equations.

CCSS.Math.Content.4.OA.A.2
Multiply or divide to solve word problems involving multiplicative comparison, e.g., by using drawings and equations with a symbol for the unknown number to represent the problem, distinguishing multiplicative comparison from additive comparison.[1]

CCSS.Math.Content.4.OA.A.3
Solve multistep word problems posed with whole numbers and having whole-number answers using the four operations, including problems in which remainders must be interpreted. Represent these problems using equations with a letter standing for the unknown quantity. Assess the reasonableness of answers using mental computation and estimation strategies including rounding.

GAIN FAMILIARITY WITH FACTORS AND MULTIPLES.

CCSS.Math.Content.4.OA.B.4
Find all factor pairs for a whole number in the range 1–100. Recognize that a whole number is a multiple of each of its factors. Determine whether a given whole number in the range 1–100 is a multiple of a given one-digit number. Determine whether a given whole number in the range 1–100 is prime or composite.

GENERATE AND ANALYZE PATTERNS.

CCSS.Math.Content.4.OA.C.5
Generate a number or shape pattern that follows a given rule. Identify apparent features of the pattern that were not explicit in the rule itself. *For example, given the rule "Add 3" and the starting number 1, generate terms in the resulting sequence and observe that the terms appear to alternate between odd and even numbers. Explain informally why the numbers will continue to alternate in this way.*

[1] Grade 4 expectations in this domain are limited to whole numbers less than or equal to 1,000,000.

Number & Operations in Base Ten & Fractions[1]

GENERALIZE PLACE VALUE UNDERSTANDING FOR MULTI-DIGIT WHOLE NUMBERS.

CCSS.Math.Content.4.NBT.A.1
Recognize that in a multi-digit whole number, a digit in one place represents ten times what it represents in the place to its right. *For example, recognize that 700 ÷ 70 = 10 by applying concepts of place value and division.*

CCSS.Math.Content.4.NBT.A.2
Read and write multi-digit whole numbers using base-ten numerals, number names, and expanded form. Compare two multi-digit numbers based on meanings of the digits in each place, using >, =, and < symbols to record the results of comparisons.

CCSS.Math.Content.4.NBT.A.3
Use place value understanding to round multi-digit whole numbers to any place.

USE PLACE VALUE UNDERSTANDING AND PROPERTIES OF OPERATIONS TO PERFORM MULTI-DIGIT ARITHMETIC.

CCSS.Math.Content.4.NBT.B.4
Fluently add and subtract multi-digit whole numbers using the standard algorithm.

CCSS.Math.Content.4.NBT.B.5
Multiply a whole number of up to four digits by a one-digit whole number, and multiply two two-digit numbers, using strategies based on place value and the properties of operations. Illustrate and explain the calculation by using equations, rectangular arrays, and/or area models.

CCSS.Math.Content.4.NBT.B.6
Find whole-number quotients and remainders with up to four-digit dividends and one-digit divisors, using strategies based on place value, the properties of operations, and/or the relationship between multiplication and division. Illustrate and explain the calculation by using equations, rectangular arrays, and/or area models.

EXTEND UNDERSTANDING OF FRACTION EQUIVALENCE AND ORDERING.

CCSS.Math.Content.4.NF.A.1
Explain why a fraction a/b is equivalent to a fraction $(n \times a)/(n \times b)$ by using visual fraction models, with attention to how the number and size of the parts differ even though the two fractions themselves are the same size. Use this principle to recognize and generate equivalent fractions.

CCSS.Math.Content.4.NF.A.2
Compare two fractions with different numerators and different denominators, e.g., by creating common denominators or numerators, or by comparing to a benchmark fraction such as 1/2. Recognize that comparisons are valid only when the two fractions refer to the same whole. Record the results of comparisons with symbols >, =, or <, and justify the conclusions, e.g., by using a visual fraction model.

BUILD FRACTIONS FROM UNIT FRACTIONS.

CCSS.Math.Content.4.NF.B.3
Understand a fraction a/b with $a > 1$ as a sum of fractions $1/b$.

- **CCSS.Math.Content.4.NF.B.3a**
  Understand addition and subtraction of fractions as joining and separating parts referring to the same whole.

- **CCSS.Math.Content.4.NF.B.3b**
  Decompose a fraction into a sum of fractions with the same denominator in more than one way, recording each decomposition by an equation. Justify decompositions, e.g., by using a visual fraction model. *Examples: 3/8 = 1/8 + 1/8 + 1/8 ; 3/8 = 1/8 + 2/8 ; 2 1/8 = 1 + 1 + 1/8 = 8/8 + 8/8 + 1/8.*

- **CCSS.Math.Content.4.NF.B.3c**
  Add and subtract mixed numbers with like denominators, e.g., by replacing each mixed number with an equivalent fraction, and/or by using properties of operations and the relationship between addition and subtraction.

- **CCSS.Math.Content.4.NF.B.3d**
  Solve word problems involving addition and subtraction of fractions referring to the same whole and having like denominators, e.g., by using visual fraction models and equations to represent the problem.

CCSS.Math.Content.4.NF.B.4
Apply and extend previous understandings of multiplication to multiply a fraction by a whole number.

- **CCSS.Math.Content.4.NF.B.4a**
  Understand a fraction a/b as a multiple of $1/b$. *For example, use a visual fraction model to represent 5/4 as the product $5 \times (1/4)$, recording the conclusion by the equation $5/4 = 5 \times (1/4)$.*

- **CCSS.Math.Content.4.NF.B.4b**
  Understand a multiple of a/b as a multiple of $1/b$, and use this understanding to multiply a fraction by a whole number. *For example, use a visual fraction model to express $3 \times (2/5)$ as $6 \times (1/5)$, recognizing this product as 6/5. (In general, $n \times (a/b) = (n \times a)/b$.)*

- **CCSS.Math.Content.4.NF.B.4c**
  Solve word problems involving multiplication of a fraction by a whole number, e.g., by using visual fraction models and equations to represent the problem. *For example, if each person at a party will eat 3/8 of a pound of roast beef, and there will be 5 people at the party, how many pounds of roast beef will be needed? Between what two whole numbers does your answer lie?*

UNDERSTAND DECIMAL NOTATION FOR FRACTIONS, AND COMPARE DECIMAL FRACTIONS.

CCSS.Math.Content.4.NF.C.5
Express a fraction with denominator 10 as an equivalent fraction with denominator 100, and use this technique to add two fractions with respective denominators 10 and 100.[2] *For example, express 3/10 as 30/100, and add 3/10 + 4/100 = 34/100.*

CCSS.Math.Content.4.NF.C.6
Use decimal notation for fractions with denominators 10 or 100. *For example, rewrite 0.62 as 62/100; describe a length as 0.62 meters; locate 0.62 on a number line diagram.*

CCSS.Math.Content.4.NF.C.7
Compare two decimals to hundredths by reasoning about their size. Recognize that comparisons are valid only when the two decimals refer to the same whole. Record the results of comparisons with the symbols >, =, or <, and justify the conclusions, e.g., by using a visual model.

---

[1] Grade 4 expectations in this domain are limited to fractions with denominators 2, 3, 4, 5, 6, 8, 10, 12, 100.

[2] Students who can generate equivalent fractions can develop strategies for adding fractions with unlike denominators in general. But addition and subtraction with unlike denominators in general is not a requirement at this grade.

Measurement & Data

SOLVE PROBLEMS INVOLVING MEASUREMENT AND CONVERSION OF MEASUREMENTS.

CCSS.Math.Content.4.MD.A.1
Know relative sizes of measurement units within one system of units including km, m, cm; kg, g; lb, oz.; l, ml; hr, min, sec. Within a single system of measurement, express measurements in a larger unit in terms of a smaller unit. Record measurement equivalents in a two-column table. *For example, know that 1 ft is 12 times as long as 1 in. Express the length of a 4 ft snake as 48 in. Generate a conversion table for feet and inches listing the number pairs (1, 12), (2, 24), (3, 36), ...*

CCSS.Math.Content.4.MD.A.2
Use the four operations to solve word problems involving distances, intervals of time, liquid volumes, masses of objects, and money, including problems involving simple fractions or decimals, and problems that require expressing measurements given in a larger unit in terms of a smaller unit. Represent measurement quantities using diagrams such as number line diagrams that feature a measurement scale.

CCSS.Math.Content.4.MD.A.3
Apply the area and perimeter formulas for rectangles in real world and mathematical problems. *For example, find the width of a rectangular room given the area of the flooring and the length, by viewing the area formula as a multiplication equation with an unknown factor.*

REPRESENT AND INTERPRET DATA.

CCSS.Math.Content.4.MD.B.4
Make a line plot to display a data set of measurements in fractions of a unit (1/2, 1/4, 1/8). Solve problems involving addition and subtraction of fractions by using information presented in line plots. *For example, from a line plot find and interpret the difference in length between the longest and shortest specimens in an insect collection.*

GEOMETRIC MEASUREMENT: UNDERSTAND CONCEPTS OF ANGLE AND MEASURE ANGLES.

CCSS.Math.Content.4.MD.C.5
Recognize angles as geometric shapes that are formed wherever two rays share a common endpoint, and understand concepts of angle measurement:

- **CCSS.Math.Content.4.MD.C.5a**
  An angle is measured with reference to a circle with its center at the common endpoint of the rays, by considering the fraction of the circular arc between the points where the two rays intersect the circle. An angle that turns through 1/360 of a circle is called a "one-degree angle," and can be used to measure angles.

- **CCSS.Math.Content.4.MD.C.5b**
  An angle that turns through n one-degree angles is said to have an angle measure of n degrees.

CCSS.Math.Content.4.MD.C.6
Measure angles in whole-number degrees using a protractor. Sketch angles of specified measure.

CCSS.Math.Content.4.MD.C.7
Recognize angle measure as additive. When an angle is decomposed into non-overlapping parts, the angle measure of the whole is the sum of the angle measures of the parts. Solve addition and subtraction problems to find unknown angles on a diagram in real world and mathematical problems, e.g., by using an equation with a symbol for the unknown angle measure.

Geometry

DRAW AND IDENTIFY LINES AND ANGLES, AND CLASSIFY SHAPES BY PROPERTIES OF THEIR LINES AND ANGLES.

CCSS.Math.Content.4.G.A.1
Draw points, lines, line segments, rays, angles (right, acute, obtuse), and perpendicular and parallel lines. Identify these in two-dimensional figures.

CCSS.Math.Content.4.G.A.2
Classify two-dimensional figures based on the presence or absence of parallel or perpendicular lines, or the presence or absence of angles of a specified size. Recognize right triangles as a category, and identify right triangles.

CCSS.Math.Content.4.G.A.3
Recognize a line of symmetry for a two-dimensional figure as a line across the figure such that the figure can be folded along the line into matching parts. Identify line-symmetric figures and draw lines of symmetry.

REFERENCES

Bambrick-Santoyo, Paul. 2010. *Driven by Data*. San Francisco: Jossey-Bass.

Chappuis, Jan. 2009. *Seven Strategies of Assessment for Learning*. Portland, OR: Educational Testing Service.

National Governors Association Center for Best Practices and Council of Chief State School Officers, *Common Core State Standards for Mathematics* (Washington, DC: Authors, 2010), accessed January 15, 2013, www.corestandards.org.

Hiatt-Michael, Diana, ed. 2001. *Promising Practices for Family Involvement in Schools*. Charlotte, NC: Information Age Publishing.

Katz, Idit, and Avi Assor. 2007. "When Choice Motivates and When It Does Not." *Educational Psychology Review* 19:429-442.

Lemov, Doug. 2010. *Teach Like a Champion*. San Francisco: Jossey-Bass.

Marzano, Robert J., Debra J. Pickering, and Jane E. Pollock. 2001. *Classroom Instruction That Works: Research-Based Strategies for Increasing Student Achievement*. Alexandria, VA: Association for Supervision and Curriculum Development.

Ozel, Serkan, Zeynee Ebrar Yetkiner, and Robert M. Capraro. 2008. "Technology in the K-12 Mathematics Classroom." *School Science and Mathematics*, 108(2), 80-85.

Petsche, Jennifer. 2011. "Engage and Excite Students with Educational Games." *Knowledge Quest*, 40(1), 42-44.

Russell, Susan Jo. 2000. "Developing Computational Fluency with Whole Numbers in the Elementary Grades." *New England Mathematics Journal*, 32(2), 40-54.

Stiggins, Rick. 2005. "From Formative Assessment to Assessment FOR Learning: A Path to Success in Standards-Based Schools." *Phi Delta Kappan*. 87(4), 324-328.

Wong, Harry K., and Rosemary T. Wong. 2005. *How to Be an Effective Teacher: The First Days of School*. Mountain View, CA: Harry K. Wong Publications Inc.